Make:

MODERN LEATHERWORK FOR MAKERS

TRADITIONAL CRAFT TECHNIQUES
MEET CNC AND 3D PRINTING

TIM DEAGAN

SAN FRANCISCO, CA

Copyright © 2017 Tim Deagan. All rights reserved.

Printed in the United States of America.

Published by
Maker Media, Inc.,
1700 Montgomery Street, Suite 240,
San Francisco, CA 94111

Maker Media books may be purchased for educational, business, or sales promotional use. Online editions are also available for most titles (*safaribooksonline.com*). For more information, contact our corporate/institutional sales department: 800-998-9938 or *corporate@oreilly.com*.

Publisher: Roger Stewart
Copy Editor: Elizabeth Campbell, Happenstance Type-O-Rama
Proofreader: Scout Festa, Happenstance Type-O-Rama
Interior Designer and Compositor: Maureen Forys, Happenstance Type-O-Rama
Cover Designer: Maureen Forys, Happenstance Type-O-Rama
Indexer: Valerie Perry, Happenstance Type-O-Rama

August 2017: First Edition

Revision History for the First Edition

2017, August 26 First Release

See oreilly.com/catalog/errata.csp?isbn=9781680453201 for release details.

Make:, Maker Shed, and Maker Faire are registered trademarks of Maker Media, Inc. The Maker Media logo is a trademark of Maker Media, Inc. *Modern Leatherwork for Makers: Traditional Craft Techniques Meet CNC and 3D Printing* and related trade dress are trademarks of Maker Media, Inc. Many of the designations used by manufacturers and sellers to distinguish their products are claimed as trademarks. Where those designations appear in this book, and Maker Media, Inc. was aware of a trademark claim, the designations have been printed in caps or initial caps. While the publisher and the author have used good faith efforts to ensure that the information and instructions contained in this work are accurate, the publisher and the author disclaim all responsibility for errors or omissions, including without limitation responsibility for damages resulting from the use of or reliance on this work. Use of the information and instructions contained in this work is at your own risk. If any code samples or other technology this work contains or describes is subject to open source licenses or the intellectual property rights of others, it is your responsibility to ensure that your use thereof complies with such licenses and/or rights.

978-1-680-45320-1

Safari® Books Online

Safari Books Online is an on-demand digital library that delivers expert content in both book and video form from the world's leading authors in technology and business. Technology professionals, software developers, web designers, and business and creative professionals use Safari Books Online as their primary resource for research, problem solving, learning, and certification training. Safari Books Online offers a range of plans and pricing for enterprise, government, education, and individuals. Members have access to thousands of books, training videos, and prepublication manuscripts in one fully searchable database from publishers like O'Reilly Media, Prentice Hall Professional, Addison-Wesley Professional, Microsoft Press, Sams, Que, Peachpit Press, Focal Press, Cisco Press, John Wiley & Sons, Syngress, Morgan Kaufmann, IBM Redbooks, Packt, Adobe Press, FT Press, Apress, Manning, New Riders, McGraw-Hill, Jones & Bartlett, Course Technology, and hundreds more. For more information about Safari Books Online, please visit us online.

How to Contact Us

Please address comments and questions to the publisher:

Maker Media
1700 Montgomery St.
Suite 240
San Francisco, CA 94111

You can send comments and questions to us by email at *books@makermedia.com*.

Maker Media unites, inspires, informs, and entertains a growing community of resourceful people who undertake amazing projects in their backyards, basements, and garages. Maker Media celebrates your right to tweak, hack, and bend any Technology to your will. The Maker Media audience continues to be a growing culture and community that believes in bettering ourselves, our environment, our educational system—our entire world. This is much more than an audience, it's a worldwide movement that Maker Media is leading. We call it the Maker Movement.

To learn more about *Make:* visit us at *makezine.com*. You can learn more about the company at the following websites:

Maker Media: *makermedia.com*

Maker Faire: *makerfaire.com*

Maker Shed: *makershed.com*

To Tracy and Pepper—
nothing makes me
prouder than the family
we've made together.

CONTENTS

About the Author v
Introduction vii

1 Understanding Leather 1
Tanning Leather 2
Understanding Hides 3

2 Fundamental Leatherworking Tasks and Tools 7
Cutting or Removing Leather 9
Joining Leather 14
Shaping and Decorating Leather 22

3 Computer Control of Leather 31
Toolchains 34
Digital Leatherworking 39

4 3D Printing Leatherworking Tools 43
Additive Manufacturing Techniques 44
The 3D Printing Toolchain 48
3D Printed Leatherworking Tools 56

5 The Fractal Journal Cover 63
Design 64
Digital Laser Fabrication 67
Leatherworking 81
The Results 85
Enhancements 85

6 The Ghouls and Gears Multi-Tool Holder 89
Design 90
Digital Laser Fabrication 93
Leatherworking 102

7 Steampunk Action-Cam Top Hat 111
Design 112
Digital Fabrication 116
Leatherworking 119
The Results 126
Enhancements 127

8 8-Bit Cell Phone Belt Case 129
Design 130
Digital Fabrication 136
Leatherworking 145
The Results 152

9 PCB Tablet Sleeve 155
Design 156
Digital Fabrication 161
Leatherworking 171
The Results 173

10 Elder Gods Belt Pouch 175
Design 176
Digital Fabrication 182
Leatherworking 187
The Results 190

11 Le Voyage dans la Lune Shoulder Bag 193
Design 194
Digital Fabrication 199
Leatherworking 203
The Results 209

A Appendix: Online Resources 211
Leather 211
Hardware 212
Software 213
Digital Fabrication 214

Index 215

ABOUT THE AUTHOR

Tim Deagan likes to make things. He casts, prints, screens, welds, brazes, bends, screws, glues, nails, and dreams in his Austin, Texas, shop. He's spent decades gathering tools based on the idea that one day he will come up with a project that has a special use for each and every one of them.

Tim likes to learn and try new things. A career troubleshooter, he designs, writes, and debugs code to pay the bills. He has worked as a stagehand, meat cutter, speechwriter, programmer, sales associate at Radio Shack, VJ, sandwich maker, computer tech support specialist, car washer, desk clerk, DBA, virtual CIO, and technical writer. He's run archeology field labs, darkrooms, produce teams, video stores, ice cream shops, consulting teams, developers, and QA teams. He's written for *Make:* magazine, *Nuts & Volts, Lotus Notes Advisor,* and *Databased Advisor* magazines.

Tim collects board games, Little Mermaid stuff, ukuleles, accordions, tools, watches, slide rules, graphic novels, art supplies, hobbies, books, gadgets, and sharp and pointy things. He owned, and escaped from owning, a 1960 Ford C-850 Young Fire Equipment fire engine (though he kept the siren). Tim paints, sketches, sculpts, quilts, sews, and works leather. Tim has climbed antenna towers, wrecked motorcycles, learned to parasail, and jumped out of perfectly good airplanes.

Tim has been, or is, a boy scout, altar boy, Red Cross disaster action team captain, volunteer firefighter, flyman, Wocista, Flipside burner, actor, Austin Mini Maker Faire flame and safety coordinator, lighting tech, ham radio operator (KC5QFG), musician, and licensed Texas flame effect operator. Tim has studied Daito Ryu Aiki Jujitsu with Sensei Rick Fine, and Tomiki Aikido with Sensei Strange.

Tim loves his wife, his daughter, his dogs, and his friends, and feels very lucky indeed to be able to write all the lists above.

INTRODUCTION

I've always loved making things. Some of the earliest things I remember getting excited about were the craft activities at the summer camps of my childhood. Of all those crafts, the one that I never stopped pursuing was leatherwork. There's something enticing about the sensory experience of leather. The look, smell, and feel are unique and appealing in a way that always makes me want to touch and handle leather items I encounter. Making my own objects out of leather is even more satisfying. It can be cut, shaped, molded, stamped, sewn, riveted, dyed, painted, carved, and tooled. Even simple objects made of leather seem to have an appeal that isn't there with other materials.

I collect tools and skills. I own 3D printers, CNC machines, vinyl cutters, a lathe, a mill, a box brake, a laser engraver/cutter, MIG/TIG/gas welding gear, an embroidery machine, a plasma cutter, and dozens, if not hundreds, of other tools. Every time I see a new tool, it seems to glow and hum with the potential of all the things it could make. I can't help but work myself into a frenzy of excitement trying to learn how to use it to create things.

Somewhere, sometime, I will manage to bring all these tools, techniques, skills, and materials that excite me together in one grand project. I'm not there yet, but I do find that crossovers between areas that seem unrelated often produce the most interesting results. This book is about my attempts to marry old-school leatherworking techniques with modern digital fabrication methods. I don't believe that either is better than the other; I believe they complement one another beautifully.

I'm not alone in this pursuit. Lots of incredible Makers have been trying out techniques like the ones we'll explore in this book. Commercial manufacturers have been using digitally controlled cutters and other tools for many years now. But I find that most people still think of their 3D printer as something that belongs in a different world than hand sewing or leather tooling. My hope is that this book provides some new perspectives on how old-school and new-school tools can work together, while sharing tips, explanations, and ideas for new directions in leatherworking.

I am incredibly lucky with all the support and help I have had that has made this book possible. My wife and daughter encourage me and support me with love and care every day. My editor, Roger Stewart, is the best in the business and has made all my writing possible. The staff at Maker Media and *Make:* magazine have supported me and made me want to learn how to make everything and to share with everyone how to come play, too. My employer, CORT Business Services, has given me the ability to have an interesting and challenging career, while being able to balance writing and making. My copy editor, Elizabeth Campbell, has made my ramblings readable, and my publicist, Gretchen Giles, has encouraged, supported, and cheered for me. I also have to thank all the people who bought my first book *Make: Fire, The Art and Science of Working with Propane* (2016, Maker Media, ISBN 978-1680450873). Without your support I wouldn't have dared to try another book.

If you take away anything from this book, I hope it's that you can color outside the lines with leatherworking. Or draw your own lines. Try new combinations of things, explore mixing and mashing ideas, parts, tools, and materials. Not everything will work together, but the delights and surprises that occur when you discover something new are worth hundreds of "Oh well, that didn't work" moments. I hope you'll visit *www.modernleatherworking.com*, where all the files for the projects in this book will be available, and you can contact me to tell me about your own.

Special Thanks: Tandy Leather

Leatherworking may be ancient in its origins and worldwide in its usage, but in the 21st century, Tandy Leather has become the single strongest source of support for people interested in learning how to work leather or purchase tools and materials. While I have no financial relationship with Tandy Leather (other than the regular money I spend at their stores), I am incredibly grateful for their support in writing this book.

Tandy has made images from their catalog, website, and stores available for this book. They have also been generous to me and thousands of other customers with their time and help on leatherworking topics and problems. This was true long before I ever considered writing a book like this.

I want to extend a special note of thanks to the manager of the Austin, Texas, store, Carmen Alexander, and the former manager (now product educator) Dennis Guerra. They, and the other employees at store #108, have always been amazingly generous to me and others who come into the store.

If you're looking for classes, materials, tools, or ideas, I hope you'll visit one of their stores or *www.tandyleather.com*. If you're near Fort Worth, Texas, visit the Al & Ann Stohlman Collection at the Museum of Leathercraft, located at the Tandy Leather Global Headquarters.

IMAGE COURTESY OF TANDY LEATHER

UNDERSTANDING LEATHER

While all-natural leather starts as some animal's skin, it's clear that there are differences between living skin and tanned leather. Given that skin will normally decompose if untreated, leather is something different. Understanding what makes leather different, and its basic qualities, helps greatly in knowing how to use leather effectively.

Skin is made of millions of tiny collagen protein fibrils (100–200 nm diameter) bound together into fibril bundles (3–6 μm), which are, in turn, bound together into fibers (30–60 μm), and then into fiber bundles about the size of a human hair (60–200 μm). (See Figure 1-1.) After the skin is removed and cleansed of hair, fat, and flesh, the collagen

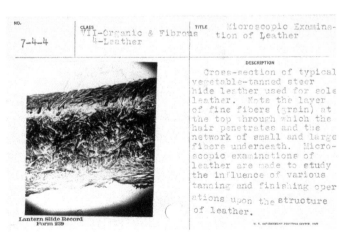

FIGURE 1-1: Collagen fibrils and fibers IMAGE COURTESY OF WIKIMEDIA COMMONS

proteins are altered so that their chemical bonds are not separated by water (aka, *hydrolysis*.) There are many methods for doing this, most of which are referred to as *tanning*. Differences in the methods used to preserve the fibers result in the distinctions between different types of leather.

TANNING LEATHER

Tanning consists of a variety of processes that permanently alter the collagens so that they do not decompose. This can be done with emulsified fat and oil, the most traditional of which is mashed-up brains mixed with water. Brain tanning is an ancient method for preserving an animal's skin, but egg yolks or even soap and oil are viable substitutes. Chrome tanning is a modern and fast method that uses chrome ions to force the water out of the collagen and then bind with the proteins. This results in most of the soft and supple leather that you encounter in clothing.

There are a number of other, less common methods, but the tanning process that results in the leather that you can easily obtain, tool, and dye is called *vegetable tanning*. Tanning derives its name from an extract of plant barks called *tannins*. Vegetable, or *veg*, tanned leather has been produced using tannins and other plant-based processes for a couple of thousand years and remains the leather of choice for most leatherworkers.

However, once the hide is tanned, it goes through several other steps before you buy it. The fibers in leather vary in their density as you move from the tightly bound grain (nearest the outer layer) to the loose fibers of the corium, which is a deeper layer of skin sometimes called the *dermis* and is closest to the flesh side of the leather. (See Figure 1-2.) As leather is prepared for sale, it is split into different grades, depending

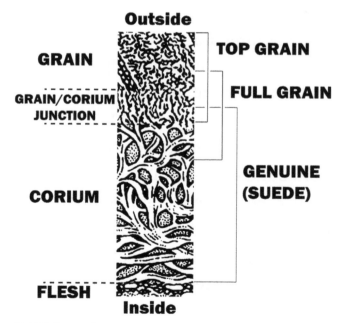

FIGURE 1-2: Grain and corium and the grades of leather

on which parts of the grain and corium are used. *Full grain* leather includes the grain and the area between the grain and corium. Since the grain closest to the hair has the densest fiber, this produces a beautiful burnished finish. The next grade is called *top grain*. Top grain shaves off the grain closest to the hair so that imperfections like scars, scrapes, or brands in the hide are eliminated. Because the densest fibers are shaved off, top grain doesn't age as well as full grain, because the fibers may start to pull apart. *Genuine leather* is the third grade, and usually starts where the grain ends. It is made up of primarily the loose fibers of the corium. Suede is a type of genuine leather. Sometimes manufacturers will buff or spray paint the top of genuine leather to make it look like a higher grade, but it doesn't wear or perform like it.

Hides that were cleaned and dried, but not tanned, are typically referred to as rawhide. Rawhide is very rigid and stiff. It can be made moldable and pliable when soaked in water, but dries, shrinks, and becomes stiff again in the molded shape. Rawhide makes great drumheads, soles for moccasins, furniture, lacing, or even shields. Because it isn't tanned, rawhide isn't considered to be leather. It's ideal for many projects, but we'll be sticking to veg tan leather for the projects in this book. (See Figure 1-3.)

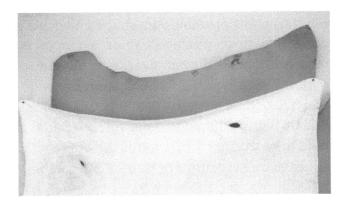

FIGURE 1-3: Rawhide compared to leather

UNDERSTANDING HIDES

Cattle hides are the most common source of leather, but skin from pretty much anything, including fish, can be used to make leather. There are several types of artificial leathers, from man-made materials like *pleather* (polyurethane) to Piñatex™, made from pineapple leaf fibers. Few of these provide the ability for you to carve, stamp, tool, and dye it like the ubiquitous veg-tanned cow hide, so we'll concentrate on cow hide for all our projects.

Veg tan leather is typically sold in thicknesses defined in ounces. One ounce is considered to represent 1/64″. While the hides are machine split, there are always some natural variations, so leather is usually listed as a range. That means that 3–4 oz leather is 3/64"–1/16" thick, while 8–10 oz leather is 1/8" to 5/32" thick. Different thicknesses are useful for different purposes. Thin leather is great for linings and lightweight projects where you'd like the leather to be flexible. Heavier leather takes tooling and stamping better and provides structural strength. (See Figure 1-4.)

It's certainly possible to buy an entire cow hide, but most people purchase leather as sectional cuts. The various parts of a hide have different thicknesses and quality, so understanding which part you want is important. Buying scraps or small, odd-shaped pieces is a cheap way to get practice leather; big projects require larger sections to provide sufficient material. (See Figure 1-5.) Because animals aren't flat, there are parts of a hide that also end up being difficult to flatten. These parts tend to be thinner and towards the outer edges of most sections of leather, but can be useful when incorporated into designs that need a curve. Leather can be reshaped and molded, but for the most part, the parts farthest from the edges are the flattest.

Leather is also sold in varying degrees of quality. Prices for the same section can be dramatically different depending on the origin and treatment of the leather. The coarseness of the grain; presence of holes, brands, or scars; details of tannage; and breed of cattle all contribute to variations in price and presentation. While using high-quality grades can result in gorgeous projects, lower grades are far more cost effective for learning how to work leather. Start off by buying the most

FIGURE 1-4: Leather thicknesses IMAGE COURTESY OF TANDY LEATHER

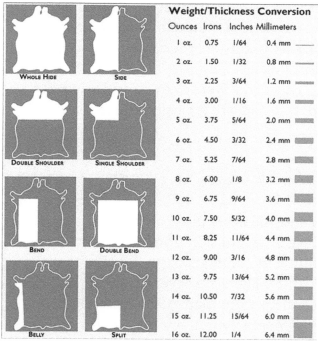

FIGURE 1-5: Parts of a hide IMAGE COURTESY OF TANDY LEATHER

affordable leather you can find, and work with it until your skills exceed its quality.

It's worth taking a moment to talk about salvaging leather from various sources. For the most part, it is difficult to reuse leather that you find in thrift shops or other sources if you're wanting to tool it or dye it. Most leather clothing is chrome tan with the exception of many leather belts. Most leather found out in the world has also had some kind of finish put on it, which makes tooling or dyeing difficult. There are, however, other leatherworking activities that do work well for recycled leather. You can add snaps, conchos, and studs on almost any leather. Cutting and sewing recycled leather is also perfectly feasible. In this book, we'll focus on projects made from fresh leather, but as long as you like the finish and color of found pieces of leather, don't give up on their project potential. (See Figure 1-6.)

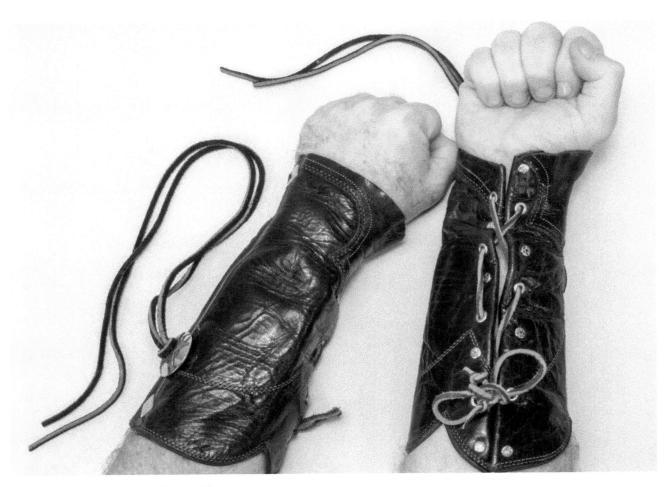

FIGURE 1-6: Salvaged leather projects

IMAGE COURTESY OF TANDY LEATHER

FUNDAMENTAL LEATHERWORKING TASKS AND TOOLS

While we will dig into the details of leatherworking in the project chapters, it's useful to survey the sets of techniques and tools that are most commonly used to work leather. Different leatherworkers will choose to use different sets of techniques. Some folks never sew; others only sew on a machine. The various preferences are championed or defended at length in forums and books. I primarily want to provide context for the leatherworking techniques we'll use later in the book.

This chapter only scratches the surface of the wealth of different tools that have been created over the years by leatherworkers, saddlers, shoemakers, armorers, artists, and hobbyists. These are the common hand tools that you are likely to encounter. Machine tools for sewing, creating lace, and working large pieces of leather are fascinating, but outside the purview of this chapter. I'm also going to leave out the huge range of dyes, paints, and finishes. They are a part of most leatherworking, and we will include their use in the project chapters rather than trying to do justice to them in a single chapter.

The single best resource for learning about leatherworking tools is a book by Al Stohlman. The late Al Stohlman and his wife, Ann, were unbelievably prolific leatherworkers throughout much of the 20th century. With more than 30 books published, the Stohlmans heavily influenced multiple generations of leatherworkers. *Leathercraft Tools* (Al Stohlman, 1984, Tandy Leather Co. ISBN 1-892214-90-3) is an absolute treasure chest of tools and techniques. I cannot recommend it highly enough for anyone interested in leatherworking. (See Figure 2-1.)

FIGURE 2-1: *Leathercraft Tools*   IMAGE COURTESY OF TANDY LEATHER

CUTTING OR REMOVING LEATHER

Whether you're cutting large pieces of leather into small pieces of leather or making holes in leather, there is a wide range of tools and techniques. Most of the tools in this section require sharpening. The specifics of sharpening all these various tools is another topic beyond the scope of this book. Nevertheless, it is essential to keep a well-honed edge on tools that work leather. You will see a significant difference in results between sharp and dull tools. Many things that seem impossible with a dull tool become easy and pleasant with a sharp one.

> ### Sharp Tools Demand Your Respect
>
> It's also worth noting that any tool that cuts, gouges, punches, or bevels leather can do the same to your skin. Exercise caution at all times when working with sharpened leather tools. You wouldn't be the first leatherworker to inadvertently dye leather with your blood, but you should still avoid injury. When working with younger leatherworkers, carefully supervise, and provide only age-appropriate sharpened implements.

KNIVES AND SCISSORS

Scissors (or *shears* if there is one large and one small hole) are a staple of leatherwork, and are used for cutting out pieces of leather from a section of hide. Heavy gauge shears, designed to provide greater cutting force, are the most common tool used for this. Small scissors, often referred to as *embroidery scissors*, are also useful for removing small fibers or tags of leather. (See Figure 2-2.)

Like many ancient crafts, leatherworking has an emblematic tool. In experienced hands, the round knife is a leatherworker's multi-tool. It can cut large sections of leather, it can make small precise cuts, and it can bevel, edge, and skive (I'll describe skiving in a moment). It's known as a *head knife*. The distinctive curved blade can be a daunting tool to new leatherworkers, but

FIGURE 2-2: Shears and scissors   IMAGE COURTESY OF TANDY LEATHER

once mastered, it frequently becomes the most used tool in a leatherworker's arsenal. (See Figure 2-3.)

A larger version with a more circular blade is known as a *round knife*. These have become less common than head knives and are less useful for operations like skiving.

Rotary knives have become an essential tool for a lot of hobbies, leatherworking among them. Available from a number of manufacturers, these knives make it very easy to cut clean, straight lines or freehand curves. (See Figure 2-4.)

Utility knives, sometimes referred to as *box cutters* and *snap-blade knives*, are also useful to cut and trim leather. These come in many sizes. The heavier duty blades are excellent for cutting very thick leather. I prefer snap-blade knives, due to the ease in maintaining a perfectly sharp tip. (See Figure 2-5.)

There are many other types of knives used in leatherwork, but they're all variations on the ones described above. Ultimately, any sharp blade that you feel confident using can be an effective leatherworking tool.

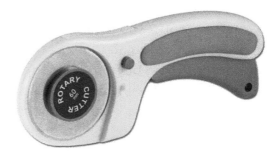

FIGURE 2-4: Rotary knife   IMAGE COURTESY OF TANDY LEATHER

FIGURE 2-3: Head knife   IMAGE COURTESY OF TANDY LEATHER

FIGURE 2-5: Snap-blade knives

SKIVING

Skiving (pronounced *sky-ving*) refers to the operation of reducing the thickness of a piece of leather. This can be done using a variety of tools, but the goal is to skive without creating divots in the leather. Skiving can occur across the breadth of the leather, such as when you would thin the end of a belt to fit through the buckle easier, or it can occur on the edges to create a taper. Skiving can be frustrating and is an easy way to discover new flavors of anger as you carve an unintended trench in a precious piece of leather. However, practice and some of the purpose-made skiving tools can make this task much easier. (See Figure 2-6.)

Sharp tools, even pressure, and a consistent angle are essential for effective skiving. When skiving with a head knife, you push the knife away from you, but the other skiving tools are pulled toward you. (See Figure 2-7.)

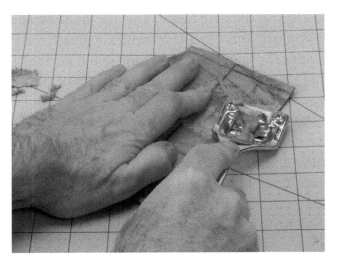

FIGURE 2-7: Skiving leather

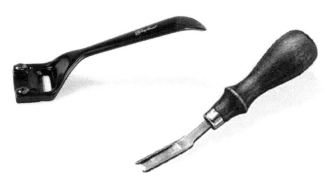

FIGURE 2-6: Skiving tools   IMAGE COURTESY OF TANDY LEATHER

SWIVEL KNIVES

When you're embossing patterns onto leather, it is necessary to make shallow cuts along many of the design lines. You can create a greater sense of depth by tamping down the leather on one or both sides of the cut. Since these lines are frequently curved or even spiraled, it is extraordinarily difficult to cut these evenly with a straight-bladed knife. The solution is to use a blade that is mounted in a swivel bearing. As you change the direction you move the knife, it changes its orientation to follow. This allows for very fluid cuts.

Traditionally, leatherworking swivel knives have a very thick blade with a wide-angle edge. They are only a few inches long and have a saddle for the index finger to rest in while the barrel is held by the thumb,

CUTTING OR REMOVING LEATHER   11

third, and fourth fingers. The standard blade is straight and used at an angle. Angled blades for filigree work are available, as are specialty blades that do things like cutting parallel lines. (See Figure 2-8.)

Swivel knives that are functionally similar are also used in CNC and vinyl cutting machines. These knives have blades of various sizes, but retain the basic model of a vertical tool with a bearing-mounted cutter. (See Figure 2-9.)

PUNCHING

Some cutting operations, such as making holes or cutting a very precise shape, are best done by punching a sharpened tool through the leather, rather than slicing or cutting. Grommets, eyelets, studs, rivets, and a number of other fixtures require holes that are as close to their outer diameter as possible. There are two basic approaches to punches: a "squeeze"-type punch that uses leverage to push the die and the anvil from both sides, and a "stamp"-type punch that relies on hammering a die through the leather with a fixed anvil underneath.

Spring or rotary punches are the most commonly used tools for punching; rotary punches are especially handy since they allow the leatherworker to rapidly change between a number of differently sized holes. These tools can be found in economy and heavy duty versions. (See Figure 2-10.)

FIGURE 2-8: Leatherworking swivel knife IMAGE COURTESY OF TANDY LEATHER

FIGURE 2-9: CNC and vinyl cutter swivel knives

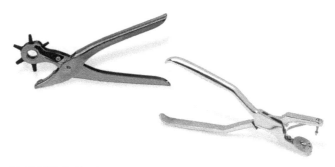

FIGURE 2-10: Spring and rotary punches IMAGE COURTESY OF TANDY LEATHER

Stamp or *drive* punches are effective for thicker leather, and also come in shapes beyond circles. These punches can create curves on corners, tapers on straps, ovals, oblong shapes, or keyhole-shaped openings for studs. These tools are generally more expensive than economy spring or rotary punches, but can provide a lifetime of use if taken care of. (See Figure 2-11.)

FIGURE 2-11: Drive punch   IMAGE COURTESY OF TANDY LEATHER

GROOVING AND GOUGING

When you need leather to bend sharply on a crease, it helps to cut a groove on the flesh side of the leather. This reduces the thickness in a narrow line and dramatically improves the quality of the fold or bend. Cutting a narrow groove on the grain side of leather to provide a guide and channel for stitching is also common. Gouging tools provide a consistent way to create these grooves.

V-gouges are the most commonly used tools for creating the larger bend grooves, but round gouges are available as well. The easiest to use are gouges that have an adjustable depth, and rest flush on the leather. Tools for creating sewing channels usually have an adjustable guide that assists in making the channel a consistent distance from an edge. (See Figure 2-12.)

FIGURE 2-12: V-gouge and stitching groover   IMAGE COURTESY OF TANDY LEATHER

Don't Hammer Metal on Metal

We will repeat this throughout this book: use a rawhide or other non-marring hammer when using metal stamps or punches. Be sure to place a thick piece of leather under the section being punched so that the sharpened end of the punch isn't hammered down onto a hard surface after passing through the target piece.

BEVELING AND EDGING

While there are numerous secondary uses they can be put to, edgers and bevellers are most commonly used to trim the sharp corners off the edges of a piece of leather, generally referred to as *beveling*. The names of these two tools are often used interchangeably, and the main distinction is that bevellers leave a flat, angled edge, while edgers can leave either a flat edge or a rounded edge. Available in a variety of widths and held at an angle, these tools can create a consistent angle along the side of a piece of leather. Beveling and edging are frequently done to improve the feel, but can also be important when creating a mitered join between two pieces of leather. These can also be used as micro-skiving tools, thinning a very narrow strip of leather. (See Figure 2-13.)

FIGURE 2-13: Different sized edging tools

JOINING LEATHER

Having covered a range of tools and methods to remove leather, we can move on to methods and tools to combine or join pieces of leather. Leather can be joined permanently, with techniques like stitching or riveting, or temporarily, with products like snaps or clasps. Lots of methods exist and all have advantages and disadvantages. A single project may use many different methods. Experimenting with different approaches to find which work best for you, technically and aesthetically, is a fun way to build your skills.

SEWING

Leather has been sewn by hand as far back as humans have worn clothes. It wasn't very long ago that leatherworkers still used a hog bristle as a needle. Hand-stitching continues to be popular because it's easy, attractive, and effective. The primary difference between stitching leather and sewing cloth is that when sewing cloth, the needle pierces the material and pulls the thread. With leather, the hole is already punched, so the needle only has to pull the thread. There are a variety of methods for creating the holes, and I'll cover the main ones below.

I'm not going to discuss machine sewing of leather. Many people prefer using a machine, but they are expensive and have a set of unique considerations that are beyond the scope of this book. It has also been my experience that once people try hand-stitching leather, they find that it is easier than they'd imagined.

There are many different types of stitches that have been created over the centuries. The most common is the saddle stitch, a manual lock stitch that is explained in depth in the Fractal Journal project. There are many other, less common types such as hidden stitches, applique stitches, and baseball stitches. Al Stohlman remains the best resource for learning more about hand-sewing leather, and I cannot recommend his book *The Art of Hand Sewing Leather* (1977, Tandy Leather Co., ISBN 1892214911) highly enough. (See Figure 2-14.)

Awls

The traditional method of poking holes for the needle involves holding a very sharp awl in one hand and pushing it through the leather just before inserting the needle. Awls come with different types of blades in various sizes, for different scale projects, and different shapes, such as curved awls for sewing tubular and adjacent pieces of leather.

To achieve an even pattern of holes, after grooving a channel for sewing, an overstitch wheel is used to mark an even set of targets for punching. As an alternative to an awl, it's also an option to use a handheld punch

FIGURE 2-14: *The Art of Hand Sewing Leather* IMAGE COURTESY OF TANDY LEATHER

that is sized especially for sewing. (See Figure 2-15.)

Some awls have a hole near the sharp point. The awl is pierced through the leather, the thread or lace is passed through the hole, and then the awl is pulled back out. Sewing awls, such as the Speedy Stitcher and Awl-For-All, look useful since they include a spool of thread and act like a manual version of a machine stitch, but in the end they are better for repairing large tarps and big projects such as banners, tenting, and custom covers than for most personal leatherworking projects. (See Figure 2-16.)

Forked Chisels

Over the last few decades, many leatherworkers have switched to using forked chisels instead of an awl. These are punched through the leather in the stitching groove, and can rapidly create a series of holes for sewing or lacing. By placing one of the tines in the last hole created from the previous punch, even spacing can be maintained. (See Figure 2-17.)

Needles

Whichever way you choose to create your holes, you'll need a method to pull the thread through them. Leather needles are heavier and have a blunter point than sewing needles. They have a relatively large eye to make it easy to thread multi-strand cord or waxed thread. Even with the heavy needle and pre-made holes, it is fairly common to break

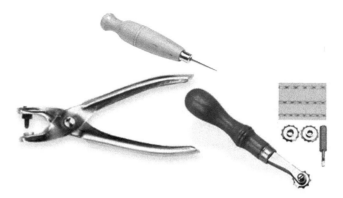

FIGURE 2-15: Awls, hand punch, and overstitch wheel IMAGE COURTESY OF TANDY LEATHER

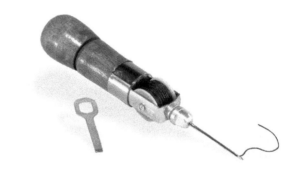

FIGURE 2-16: Sewing awl IMAGE COURTESY OF TANDY LEATHER

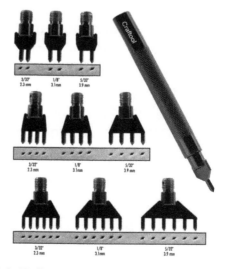

FIGURE 2-17: Forked chisels IMAGE COURTESY OF TANDY LEATHER

needles when sewing leather, so most leatherworkers buy these in sets of ten or more. (See Figure 2-18.)

SCREWS, GROMMETS, AND RIVETS

There are many reasons why sewing may not be the best choice when joining pieces of leather. Luckily, there are other methods available to leatherworkers. These are generally methods that hold leather together with metal that is either screwed, bent, or hammered in a way that pins the two pieces together. Much like sewing, this is typically done using a pre-punched hole.

Screws are fairly self-explanatory. The piece on one side has a male thread and the piece for the other side has a female thread. One is screwed into the other. They are easy to assemble, but you run the risk of them unscrewing themselves at an inopportune time. They are best for decorative rather than structural use. Screws generally have a design element, typically either the shape or the pattern. (See Figure 2-19.)

Grommets provide a reinforced hole. One side has a ring that circles the hole and a post that goes through it. The other side has a washer that forms a second ring. The post is deformed to pin the washer in place. Grommets range in size from tiny to very large (grommets used on tarps are

FIGURE 2-18: Leather-stitching needle   IMAGE COURTESY OF TANDY LEATHER

FIGURE 2-19: Screw post   IMAGE COURTESY OF TANDY LEATHER

essentially the same as grommets used on leather).

Eyelets are similar to grommets, but have only one piece, the ring and post. When installed, the post deforms directly against the back side. From the top side, an eyelet looks very much like a grommet. Grommets are stronger and better for extended use, but even a small eyelet can improve a project considerably.

Both grommets and eyelets provide a dramatic enhancement to the strength and durability of a hole. They are frequently used on a single piece of leather just for this strengthening ability. Unprotected holes in leather that get a lot of wear, perhaps being pulled by a buckle or tugged by string, are often the first thing to wear out on a leather project. Placing a grommet or eyelet is a cheap and easy way to avoid this problem. (See Figure 2-20.)

FIGURE 2-20: Eyelets and large grommets

Rivets come in a variety of styles. We'll use a couple of different types in the projects to get more familiar with them, but we'll go over the main types here. The original type of rivet, still commonly used today, is a thin rod of metal (frequently copper) that has been hammered (also called being *peened*) into a mushroom shape. Most rivets today come with a pre-flattened side so that you only have to peen one end. The rivet post is first passed through a hole in the pieces to be joined. Next, a washer is placed around the post on the back side. Then, the post is cut to the proper length and peened down to hold the washer in place, pinning the pieces of leather between them. (See Figure 2-21.)

Consistently peening rivets can take some practice. Many people prefer to use newer alternative approaches. While these offer convenience, they are generally not as strong as a well-set copper rivet.

Tubular rivets are a single piece joining tool. After passing the rivet's hollow post through a hole in the leather, a setting tool is used to hammer down the post. The setting tool splits the post into a number of thin sections (I'll refer to them as *tines*) that are guided to curl down and pinch the leather in place. Tubular rivets are easy to be successful with, but can fail over time if the tines become stressed enough to crack off. (See Figure 2-22.)

Another type of rivet, known as the Rapid Rivet, is very popular. It consists of a base with a post and a cap. The post is pushed through the leather, ideally leaving about 1/8" so the cap can be placed on top. The setting tool is hammered to set the cap securely on the end of the post. These rivets are also available as a "double cap"

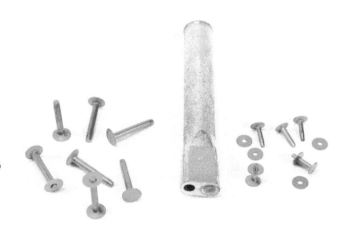

FIGURE 2-21: Copper rivets with washers

FIGURE 2-22: Tubular rivet   IMAGE COURTESY OF TANDY LEATHER

rivet that uses a concave anvil on the bottom to hold the cap on the back side. (See Figure 2-23.)

CLOSURES

Snaps

While you *could* choose to rivet the cover of a bag in place, the value of the durability gained would be diminished by the inability to open and close the bag at will. Many projects require a means to secure a cover, and many types of closures have been created to help.

Snaps are a very popular closure for good reasons. They provide a positive tactile, and sometimes audible, feedback; they can generally be opened with one hand but resist accidental opening; and they come in a variety of colors, sizes, and designs. Most snaps require holes in the leather on each of the two pieces to be joined. The cap and socket attach to the top piece, and the stud and eyelet attach to the bottom. The term *eyelet* is used in this case to denote something different than we were describing with grommets. (See Figure 2-24.)

Buckles and Rings

Whether it's wide like a belt or narrow like a watchband, there are a variety of options for securing a leather strap. The two main ways are to pierce holes in the strap and use a buckle with a prong to hold it, or run it through a pair of rings so that tension holds it tight.

Buckles come in a variety of configurations: center bar, roller, two prong, double bar, even the Conway buckle that has no moving parts. The addition of different sizes, finishes, and decoration provide a vast array of different buckles to choose from. (See Figure 2-25.)

Rings can be used to retain straps or belts or, when doubled, to secure them. Rings are generally circular or D-shaped, but are

FIGURE 2-23: Rapid Rivet IMAGE COURTESY OF TANDY LEATHER

FIGURE 2-24: Line 20 snap IMAGE COURTESY OF TANDY LEATHER

FIGURE 2-25: Buckles IMAGE COURTESY OF TANDY LEATHER

sometimes built into a plate to allow multiple straps or belts to connect to a central spot. (See Figure 2-26.)

Hooks and Clasps

Like buckles, hooks and clasps have evolved into a staggering array of different styles and looks. Clasps are usually metal or plastic. Most come in two pieces, but some, like watch clasps, can be a single piece. Clasps vary in how securely they connect things. A bag clasp is usually fairly strong, while a swing clasp might be better suited for something intended not to move much. (See Figure 2-27.)

Buttons and Studs

Buttons, in the sense of the ones used on most clothing, are uncommon in leatherworking projects. But an older style of button, the toggle button, can be a great addition to a leather project. Instead of a flat circle, the toggle is a cylinder that is attached at its centerline. It slips sideways through a loop of material and buttons securely, but is easy to undo. (See Figure 2-28.)

Studs used for closures, as opposed to purely decorative studs, are usually a sphere mounted on a stem. The stem generally has

FIGURE 2-26: Rings   IMAGE COURTESY OF TANDY LEATHER

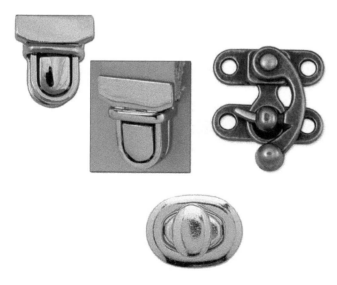

FIGURE 2-27: Bag, swing, and other clasps   IMAGE COURTESY OF TANDY LEATHER

FIGURE 2-28: Toggle button

a flat base and uses a screw mount to attach to the leather. The stud is lined up with a hole just slightly smaller than the diameter of the sphere that has a slit. The stud is pushed through the hole, using the slit to help fit the sphere through, and then wraps around the stem. Studs can provide a surprisingly strong closure and are easy to open and close with one hand. (See Figure 2-29.)

Magnets

Magnets have also started showing up as a closure option. They can be sewn into a project or attached with tines or a rivet back. While magnets aren't great for securing something strongly, they have the advantage of being very easy to close, sometimes even closing by themselves. Magnets could be incorporated into a design as a "quick close" method, or with a stronger clasp or hook as a "secure close" method. (See Figure 2-30.)

GLUE

Glue has been used for centuries to attach pieces of leather, and remains an important tool for leatherworkers. A huge range of leather glue products are available, but rubber cement, cyanoacrylate (Super Glue), shoe glue, and other products can be useful. Glue can be used to hold items while sewing, or as the only method of permanently attaching leather. The "proper" or "best" type of glue is a topic of heated conversation on various internet forums. Many people have strong opinions. It's important to experiment to determine how the type and thickness of the leather you're working with behaves. How well you can tolerate smells, holding and drying time, and other factors are all important for deciding which glues you might want to use. (See Figure 2-31.)

FIGURE 2-29: Stud   IMAGE COURTESY OF TANDY LEATHER

FIGURE 2-30: Magnet closures

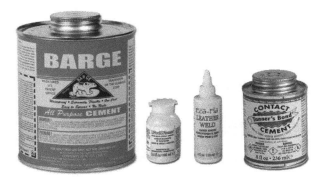

FIGURE 2-31: Various leather glues   IMAGE COURTESY OF TANDY LEATHER

SHAPING AND DECORATING LEATHER

Beyond the structural aspects of cutting or joining leather, there are many ways to decorate it. The leather can be shaped to display patterns or images, or it can be adorned with metal, rhinestones, or ornamental items like conchos.

FIGURE 2-32: Spots   IMAGE COURTESY OF TANDY LEATHER

SPOTS AND CONCHOS

Patterns of spots (sometimes referred to as studs) are a familiar site on biker jackets and other apparel. These may be metal or have glass set in them for a jeweled effect. Usually these have small tines on them that are stuck in the leather and bent to hold the item in place. Some may have screw backs and mount to a hole. (See Figure 2-32.)

Conchos generally have a screw or rivet back and attach to a hole in the leather. Some have slots and are tied to the leather with laces. They usually have some design or pattern on them and are larger than spots. (See Figure 2-33.)

We'll design our own items to attach to leather with a 3D printer in Chapter 7, "Steampunk Action-Cam Top Hat."

FIGURE 2-33: Some of the author's conchos

MOLDING

Wet leather can be stretched and molded into surprisingly complex shapes. When the leather dries, it will retain the shape it was molded around. There aren't any specific tools that are required, but it does help to have a few things to assist in the process.

If you're making a case for a cell phone and you want the leather molded around the shape of the phone, it's unlikely that you'll be happy wrapping the phone in wet leather and leaving it there until it dries. Creating a model of the phone or other object in a moisture-safe material is the solution. Carving the model in wood is a traditional approach; we'll use a 3D printer to make ours.

It also helps to have something to press around sharp bends and stretch it to the desired shape. This can be almost anything with a smooth finish (to avoid marring the leather). Bone, sanded wood, plastic, and metal can all be used if the implement fits the profile you're wanting.

The leather can be held in place with clips or pins. It is a frustrating surprise to discover that steel or iron clips, nails, tacks, and pins have stained your leather an ugly black. Use brass pins, tacks, or nails and wooden or plastic clips. It also helps to cut small pieces of thin leather to use over the jaws of clips so that they don't transfer their ridges to the receptive leather. (See Figure 2-34.)

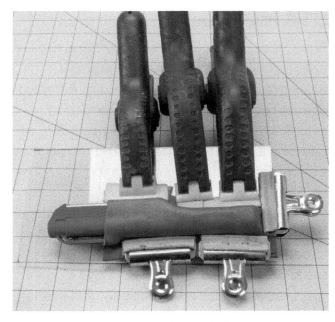

FIGURE 2-34: Molding leather with clips

TOOLING AND CARVING

Carving or tooling leather shapes the leather into a low-relief canvas that can represent almost anything. Oddly, the term *carving*, when applied to leather, generally doesn't mean cutting pieces off. The term is used to describe operations that selectively depress leather to achieve effects, usually in a free-hand manner. Using shaped stamps to pattern leather is mostly referred to as *tooling* the leather. The two techniques can be used independently or in combination.

CARVING

Leather carving tools are similar to the tools used by potters and sculptors working in clay. Small spoon-shaped tools depress and smooth, tools with spherical beads on the end create lines, and spade-tip tools create hard corners or angles. With a very basic set of carving tools, usually in conjunction with a swivel knife, a leatherworker can create bas-relief images with tremendous detail. (See Figure 2-35.)

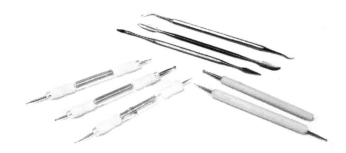

FIGURE 2-35: Leather carving tools

SHAPING AND DECORATING LEATHER

STAMPING

From very simple tools made by filing the heads of nails to shockingly expensive high-grade steel stamps, a world of different shapes that can be pounded into leather is available to leatherworkers. Stamps come in patterns that can be repeated and motifs that tie a variety of different stamps together. The technique of tooling leather generally combines lines cut with the swivel knife, shaders that depress one or both sides of those lines to add depth, and patterns or textures that bring detail to life.

As we've discussed previously, a rawhide or other non-metal hammer should be used when stamping leather. A clean, smooth, solid platform underneath the leather is also essential for getting good results. A block of granite 1" thick or more is the best surface, though I've had good results with cementing two ½" granite tiles together to make a more affordable surface.

Collecting stamps is fun, but dangerous for the obsessive. Hundreds of different stamps have been created over the years, and it's easy (at least for me) to find yourself constantly in search of new ones. We'll use digital fabrication tools in the project chapters to create our own custom stamps, opening up a whole new world of opportunities. (See Figure 2-36.)

FIGURE 2-36: Leather stamps   IMAGE COURTESY OF TANDY LEATHER

Hand-Stitching Leather

Though its history dates back to the dawn of civilization, leatherworking remains an enjoyable and useful skill, even in the age of 3D printers. While you can spend a lifetime learning the deeper intricacies of leather, the basics are easy enough for anyone to pick up. Among the most useful of these skills is the ability to sew pieces of leather together. The process is similar to sewing cloth, but has some significant differences. In this skill builder, we'll learn how to hand-sew leather using the saddle stitch.

Hand-sewing leather may seem daunting to many Makers, but it's inexpensive, very strong, and less work than you might imagine. The saddle stitch is actually more durable than a machine stitch. When a machine stitch breaks, the entire piece will quickly unravel. When a saddle stitch breaks, the threads bind each other in place. (See Figure 2-37.)

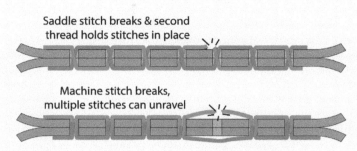

FIGURE 2-37: The manual saddle stitch vs. machine stitching

Stitching needles are heavier, longer, and duller, and have a larger eye than regular sewing needles. Unlike when sewing cloth, the needle is not intended to create its own hole. A hole is punched through the leather by an awl or chisel, then the needle is pushed through. We'll use two needles, one on each end of waxed thread. This thread is much heavier and stronger than cloth thread and typically made from multiple cords of strong linen or synthetic material. A small lump of beeswax will help bind the thread. (See Figure 2-38.)

FIGURE 2-38: Basic hand-sewing leather tools

Cut a length of thread the distance between your outstretched hands. For big projects, double that. Pass the end of the thread through the eye of the needle, then pierce the tip through the thread about 3" from the end. Personally, I always pierce the thread twice. (See Figure 2-39.)

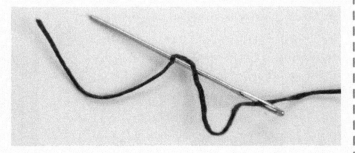

FIGURE 2-39: Piercing the thread with the needle

SHAPING AND DECORATING LEATHER

Slide the thread down the needle until it passes the eye, then draw it tight (see Figure 2-40). Rub the beeswax along the splice and roll it tight between your fingers. Do the same on the other end of the thread with the second needle.

FIGURE 2-40: Pull the thread to tighten the splice.

Next, we will prepare the leather. We need to score a line that is the same distance from the edge of the leather as the thickness of the two pieces of leather being sewn. There are fancy tools for doing this, but you can use a pair of scissors like a compass to accomplish the same thing. (See Figure 2-41.)

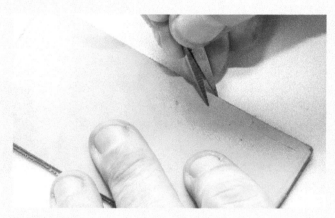

FIGURE 2-41: Mark a line for sewing.

The distance between the holes varies with the intended use, thickness of the thread, and weight of the leather. If you're using an awl, an overstitch wheel is the best way to mark the locations. While using an awl is old-school cool, chisel forks have become much more popular. Place the two pieces of leather together in the position you want to sew them. Set them on a piece of smooth wood with a thick piece of leather on top that you don't mind damaging. Hold the fork along the marked line and use a soft-headed hammer to punch it through the leather. Drive the fork all the way through the two pieces being punched. Pull the fork out, set the first prong in the last hole, and punch the next section (see Figure 2-42.) Continue until holes are punched along the length to be sewn.

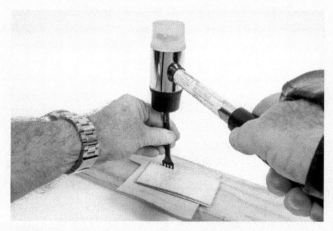

FIGURE 2-42: Punching the holes

The traditional way to hold leather while sewing is in a stitching horse (or the smaller version, called a *pony*). (See Figure 2-43.)

You can also hold it between your knees, in a soft-jawed vise, or in any manner that will leave your hands free (two needles need two hands). You can even let it dangle while you sew, though this will not help your speed or consistency. However you decide to hold it, pass one of the needles through the hole where you want to start and pull it until the workpiece is in the middle of the thread. (See Figure 2-44.)

Take the needle that will be on the back side of your work and pass it back through the next hole closest to you. We will stitch toward ourselves. Pull 2" of thread through the hole. Take the needle on the front side and push the tip just through the hole in front of the thread that's coming through. We always place the needle on the front side in front of the thread from the needle on the back side. (See Figure 2-45.)

FIGURE 2-43: My home-built stitching pony

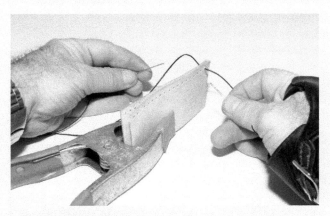

FIGURE 2-44: Ready to start sewing

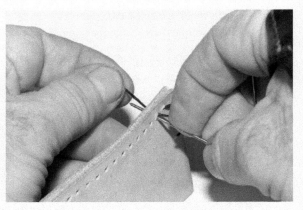

FIGURE 2-45: Placing the needle on the front side in the hole

Before you pull the needle on the front side through the hole, we need to make sure it didn't pierce the incoming thread. If that happens, the stitch will have to be cut and you have to start over (or learn the advanced skill of dealing with a pierced thread). We can avoid this by pulling the incoming thread back through the hole as we push the needle on the front side into the hole. When the needle on the front side is almost all the way through, we can stop pulling the incoming thread. Then, we take a needle in each hand and pull evenly until the stitch tightens. (See Figure 2-46.)

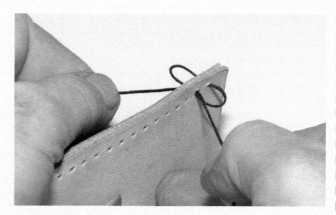

FIGURE 2-46: Pulling the stitch tight

Continue this sequence along the row of holes until you come to the end. To finish and secure the threads, we'll *back-stitch* for two holes. This means that we will change direction and stitch over the last two stitches. (See Figure 2-47.)

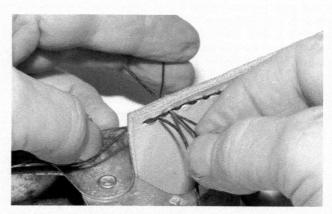

FIGURE 2-47: Back-stitching to finish off

The needles will be harder to get through the holes that already have thread in them. I generally end up using needle-nose pliers to pull the needle through. Be careful doing this, since it makes it easy to break the needle, almost always at the eye. You can avoid this by carefully pulling straight through and not putting any side force on the needle. Breaking off a needle can be dealt with if there's enough line to thread a new needle and keep going. Otherwise, you'll have to back-stitch as much as you can with the other needle, and hope. (See Figure 2-48.)

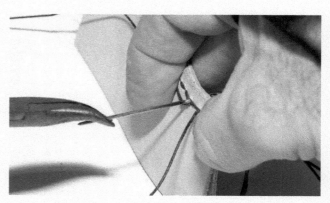

FIGURE 2-48: Pulling a needle through

Once you're finished, use small scissors or a utility knife to cut the remaining threads as close to the leather as possible. Many leatherworkers will have used a special tool to gouge a shallow trough along the line of holes and, when finished sewing, use a hammer to tap the stitches down into the trough. This keeps them from experiencing as much wear, and makes them last longer.

This is only the most basic of stitches. If you'd like to learn more about hand-sewing leather, the master reference is Al Stohlman's *The Art of Hand Sewing Leather*. This excellent instruction book has taught tens of thousands of leatherworkers basic and advanced techniques. (See Figure 2-14.)

With a little practice, hand-sewing becomes a fast, easy, and fun way to make anything from a wallet to a saddle. Give it a try and discover a whole new world of leatherworking.

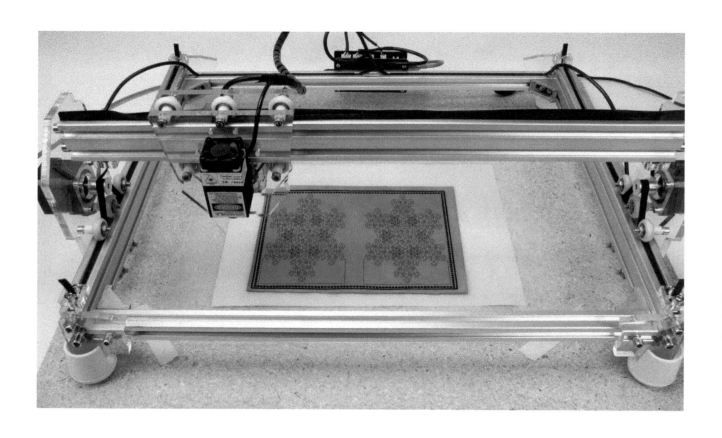

COMPUTER CONTROL OF LEATHER

Though the fundamentals of machine automation go back to the Jacquard loom in the beginning of the 19th century, John Parsons was awarded patent #2,820,187 in 1958 for "Motor Controlled Apparatus for Positioning Machine Tool" and invented numeric control (NC) as we know it today. Early NC machines used punched tape to record sequences of operations. When computers began directly controlling the machines, computer numeric control (CNC) became a powerful new force in manufacturing. (See Figure 3-1.)

FIGURE 3-1: Paper tape program IMAGE COURTESY OF WIKIMEDIA COMMONS

Commercial operators have been using digitally controlled tools to produce leather products for decades. Hobbyists have been doing so for nearly as long, but there is little shared about this work compared to other topics in CNC, 3D printing, lasers, and so on. Digital fabrication tools have become dramatically more available to Makers, crafters, and DIY enthusiasts with the advent of the internet. As the market expands and increasing numbers of people have access to vinyl cutters, 3D printers, CNC machines, laser engravers, and cutters, the urge to explore additional materials grows.

We could rightly apply the term *CNC* to any of the digital fabrication tools I've listed above. A 3D printer is a CNC plastic oozer. A vinyl cutter is a CNC drag knife. We tend to restrict the common use of the term *CNC* to describe metal or woodworking tools that have a spindle or router and a gantry design. They are generally *subtractive manufacturing* tools, in that they cut material away (as opposed to a 3D printer, which is an *additive manufacturing* tool). But this is a loose constraint, and we might refer to a CNC plasma cutter, CNC lathe, or CNC hot wire foam cutter just as easily. (See Figure 3-2.)

In this book, I'll use the term *digital fabrication* to refer to the range of

FIGURE 3-2: CNC router in a workshop   IMAGE COURTESY OF WIKIMEDIA COMMONS

different tools that perform additive or subtractive manufacturing and are controlled by a computer. I'll use the term *CNC* to refer to a Cartesian (X, Y, Z axes) machine with a spindle or router that can take different bits. This could be a DIY machine, ShopBot, Tormach, Shapeoko, or, for most of the pictures in this book, an X-Carve from Inventables. There is no practical difference between any of these in regard to the projects in this book. If the tool can use milling cutters, V cutters, or a drag knife, it will work great.

TOOLCHAINS

With a computer comes software. The software associated with digital fabrication is a huge collection of programs that do a host of different things. Nevertheless, there is a consistent set of things we expect the computer to do when engaging in digital fabrication, and the set of one or more programs that do these things is referred to as the *toolchain*.

Different problems may require alternative approaches, but the common toolchain involved in digital fabrication involves the following:

1. Construction of an image or model
2. Generation of a toolpath
3. Execution of a toolpath
4. Conversion of a toolpath to step or direction commands
5. Conversion of the step or direction commands to a pulsetrain

CONSTRUCTION OF AN IMAGE OR MODEL

We must start with some picture or 3D model of what we want the machine to create. In the case of an image, we might use any of hundreds of graphic design programs. The image could be a raster bitmap consisting of pixels or a vector image consisting of descriptions of lines and curves. Raster images are typically fabricated a line at a time, scanning through the rows of pixels. Vector images are fabricated by moving the tool along the lines and curves in the file.

Machining can occur in 2D, 2.5D, or 3D modes. Machining in 2D means that all cuts or features are at the same depth. Water jet cutters, laser cutters, and stencil cutters are typically 2D operations. Machining in 2.5D refers to projects where you can have different depths for the tool head, but they're all flat. The Z axis doesn't change while X and Y axes are moving. Machining in 3D involves moving the X, Y, and Z axes at the same time. (See Figure 3-3.)

In this book, I'll refer to the free program Inkscape (inkscape.org) when I'm discussing the creation of vector files. There are many other excellent programs; perhaps the best known is Adobe Illustrator. In keeping with my enthusiasm for open-source and free programs I'll refer to the GNU Image Manipulation Program (GIMP; www.gimp.org) when discussing raster image work. Adobe Photoshop is the most popular commercial equivalent, though there are hundreds of drawing and graphic programs out there that you might choose to use.

Computer assisted design (CAD) programs are typically used to build 3D models (and many 2D models). There are many powerful CAD tools out there, and I'll list a

FIGURE 3-3: Machining in 2D, 2.5D, and 3D

number of them in the appendix. For this book, I'll refer to Autodesk® Fusion 360™. This phenomenally powerful tool is, as of this writing, free for hobbyists, students, and small businesses making less than $100,000 per year. The tasks I'm performing could be done in any 3D CAD program, but Fusion 360 is a powerful tool to have in your arsenal. (See Figure 3-4.)

Common raster image formats include JPG, BMP, GIF, and PNG. Common vector formats are SVG and DXF. DXF (Drawing Exchange Format) is the most common format for moving vector files between programs. In this book, we'll stick to PNG, SVG, and DXF files for images. There are dozens of possible formats for 3D models. We'll primarily use STL files since they've become the *de facto* standard with the rise of 3D printing.

GENERATION OF A TOOLPATH

A toolpath is just what it sounds like: the path you want the tool to follow. But that's where the simplicity ends. Determining a toolpath can be a devilishly complicated activity. Toolpaths need to take into account things like the width of the tool, the order of operation for cuts or movements, whether the activity is on the inside or outside of a model, the speed of the tool and the rate of movement, the optimal set of movements to accomplish the goal in the shortest period, whether there should be tabs left in place so the material doesn't go flying off the table, and hundreds of other practical details.

FIGURE 3-4: 3D model

Which things are addressed in the toolpath is dependent on what kind of tool you're using and what operation you're asking of it. Models intended for 3D printing are run through a "slicer" program that breaks the model into layers and defines the movement of the extruder on each layer. Slicers have to calculate whether the tool is creating an external shell or generating infill. They have to decide if subsequent layers are going to need additional support on the layer under construction, and they have to determine if the current layer's movement is an unsupported "bridge" with nothing on layers below it. The slicer changes the rate of extrusion and the temperature of the hot end and bed based on dozens of considerations.

CNC toolpaths must take into account the cutting tool's width, desired speed, feed rate, and profile. They must make sure to move the tool in the appropriate direction for the cut desired. When cutting corners, the tool may need to move in a surprising dance to result in the edge that was intended. (See Figure 3-5.)

Toolpath creation can be done by an independent program, or the functionality can be bundled into a program that provides multiple toolchain features. Even for a given tool, different toolchains can be used. (See Figure 3-6.)

There are many languages that have been developed to describe toolpaths. The predominant one is G-code. G-code is a (relatively)

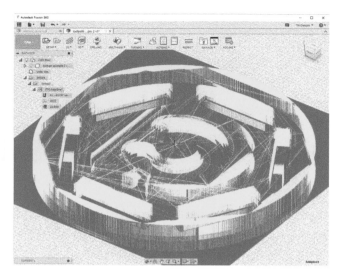

FIGURE 3-5: CNC toolpath

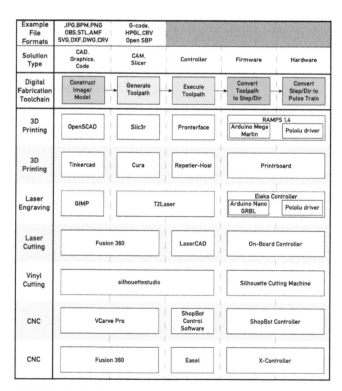

FIGURE 3-6: Comparative toolchains

36 COMPUTER CONTROL OF LEATHER

simplistic language that provides commands to move a tool around, as well as specialty commands for things like turning fans on or off, applying lubricants, or returning diagnostic information. G-code is used by many digital fabrication tools, but it has a wide range of implementations. G-code for a 3D printer resembles G-code for a CNC router, but there are many tool-specific differences in some of the commands. Some manufacturers have even gone beyond G-code, extending it to include logical programming features like loops and if-then statements.

EXECUTION OF THE TOOLPATH

Many tools rely on an independent controller program to send the G-code to the machine. Some have integrated controller software—3D printers are controlled by tools like Pronterface or Repetier-Host. CNC tools can be controlled by G-code senders like Universal G-code Sender, SB3, Easel, or any of dozens of other tools. My vinyl cutter has an all-in-one tool software package that does everything from construct the image to control the tool.

The controller typically listens to the return messages from the tool to detect end-stops, temperatures, or other instrumented values, as well as status. The return messages define the cadence at which the commands are sent by providing acknowledgement that the previous command was completed.

CONVERT THE TOOLPATH TO STEP AND DIRECTION COMMANDS

Most digital fabrication tools use stepper motors to move the tool. *Stepper motors* are different than regular motors in that instead of just spinning when power is applied, they can be moved a very specific number of degrees. This movement is referred to as a *step* and it can be clockwise or counterclockwise. A common step angle is 1.8°, which means a full revolution would take 200 steps to complete (1.8° × 200 = 360°, which is a full circle). Belts or a leadscrew convert this circular motion into linear motion, turning the degrees of a step's revolution into a distance travelled on an axis.

There are alternatives to steppers. Servos are the most common, but they have limited use in most hobbyist- and Maker-level tools, so I'm going to avoid discussing them in this book.

Step and direction commands can be issued millions of times during digital fabrication. Each command is a single bit; yes or no for step, and forward or reverse for direction. When issued to multiple motors driving different axes and combined with commands to turn extruders or cut heads on or off, as well as messages from sensors, they compose the backbone of digital fabrication activities.

CONVERSION OF THE STEP AND DIRECTION COMMANDS TO A PULSE TRAIN

Even with low-level, single-bit step and direction commands, we're not at the level where we control the motors directly. Stepper motors come in a variety of configurations. Among the differences is the number of wires used to control the motor. Many digital fabrication tools use four-wire steppers. Ultimately, the number of wires reflects how the coils in the motor are wired. Four-wire motors are very simple: there are two coils and the wires are their leads. (See Figure 3-7.)

FIGURE 3-7: Four-wire stepper motors

To move the motor one step, the coils must be energized in a specific sequence. This typically occurs with a series of square wave pulses. Different motor configurations require different pulse trains. For the simple four-wire motor we're discussing (known as a bipolar motor), the *full step* pulse train has four steps and the *half step* has eight steps. (See Figure 3-8.)

Conversion of the step and direction commands to a pulse train can occur in several ways. Some devices will use a microcontroller to convert the step and direction signals; others will use a dedicated chip. Because many motors run at relatively high currents, a motor driver chip is usually the last stage before the motor itself. (See Figure 3-9.)

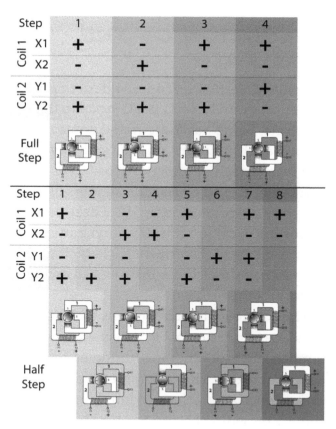

FIGURE 3-8: Four-wire bipolar half and full step pulse trains

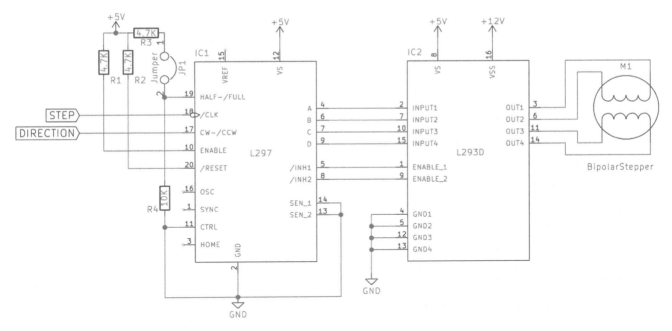

FIGURE 3-9: Stepper motor driver schematic

DIGITAL LEATHERWORKING

The processes described above for digital fabrication can be used with a wide range of materials. Wood, metal, and synthetic materials such as Delrin® are commonly used in CNC operations. Leather is not unknown, but fewer than one percent of CNC hobbyists surveyed say they cut leather this way.

Commercial manufacturers use CNC machines to cut leather, but that's typically on large tables and most frequently involves chrome tanned leather for clothing. Cutting or tooling smaller veg tanned leather pieces on home machines is less common, but still extremely useful.

Other digital fabrication techniques also offer interesting opportunities for leatherworking. As we'll see in the project chapters, not only can we CNC leather, we can engrave it with lasers, use vinyl stencils to do dye masking, and create tools, molds, and conchos on 3D printers.

UNIQUE PROPERTIES OF LEATHER

Leather differs from most of the materials commonly used with digital fabrication tools in a number of ways. These differences can be daunting if you're unfamiliar with leather, but are relatively easy to overcome with some consideration.

Thickness

For operators used to pulling out their calipers and measuring their materials in thousands of an inch, leather is frustrating. A section of leather can vary in thickness by 10 to 20 percent at different points. Hair follicles, brands, and scars also contribute to the variable nature of leather. It's possible to plane leather to make it more consistent, but I'm of the school of thought that embraces leather's quirks and variances. If this doesn't work for you, buying higher-quality leather is the best way to get better consistency.

Thickness isn't an issue for many of the digital fabrication techniques we'll use, since we'll effectively be using 2.5D operations on the leather. Where this isn't viable, we should restrict ourselves to cutting depth ranges that take the variance in thickness into account, treating the entire piece as if it were the thickness of the narrowest point.

Laser cutting operations can also be impacted by variances in thickness. The best way to handle this is to set the power to the minimum level sufficient to cut the average thickness and then use additional passes of the laser to cut the thickest spots.

Hydrophilicity

Okay, in fairness, I could have just titled this section, "Water," but how often do you get the chance to say "Hydrophilicity?" The word basically means that leather readily absorbs water. Veg tan leather soaks up water and other liquids, expanding and becoming flexible. When wet, it can be stretched and molded into complicated shapes. If allowed to dry while being molded, it will retain the shape. (See Figure 3-10.)

This is, in general, a useful property. It allows leather to be formed into shapes and profiles that would be extraordinarily difficult to form through cutting and stitching. We'll be using this property in some of the project chapters to shape leather. It is frustrating, however, to make measured cuts on leather when it's wet, since the leather contracts as it dries. Worse yet, it does not necessarily contract evenly.

In the chapters where we're cutting leather with a laser, I discuss the merits of cutting leather wet and dry. If dimensional accuracy is of prime importance, then cutting dry is necessary. I prefer cutting wet

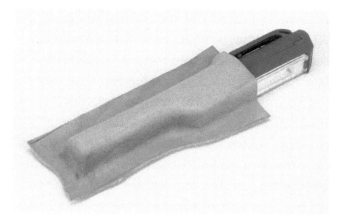

FIGURE 3-10: Molded leather

when possible to limit the charring and smell created by burning the leather with the laser.

The tendency to absorb water can also be a challenge when applying dyes or finishes. I'll discuss why I find an airbrush indispensable in the journal cover project. Suffice to say that putting a brush or sponge loaded with dye onto leather causes it to suck the liquid up and cause blotches.

Surprises

Lastly, I must confess that leather is often full of surprises. Unseen creases may reveal themselves after a piece has been cut. There can occasionally be areas that don't take dye as well as other areas. Again, higher-quality leather tends to reduce these kinds of issues, but I prefer to work with these inconsistencies when possible.

It's not always possible to include these inconsistencies in a project. Sometimes you want to do your best to avoid them. Other than buying expensive cuts of leather, the best way to avoid surprises is to avoid using the very edge of a section where possible. Use the edge pieces for dye tests and cutting practice. Make your rough cut from an area closer to the center or main area of clean leather. Most of the inconsistencies tend to be on the outer areas of pieces.

Another surprise that's easy to avoid is that leather gets sunburned. When storing your veg tan leather sections, be sure that there isn't any sunlight (or light with a high UV content) that will hit it. Leather exposed to sun darkens considerably. (See Figure 3-11.)

This is a process that some leatherworkers use consciously to naturally age and darken the leather. The process will occur with all veg tan leather exposed to UV. Most people find this a "mellowing" that is part of leather's appeal. But it's very frustrating when you reach for the expensive section of leather you bought and discover that it's half darkened where the sun shined on one side.

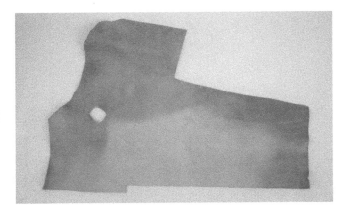

FIGURE 3-11: Suntanned leather

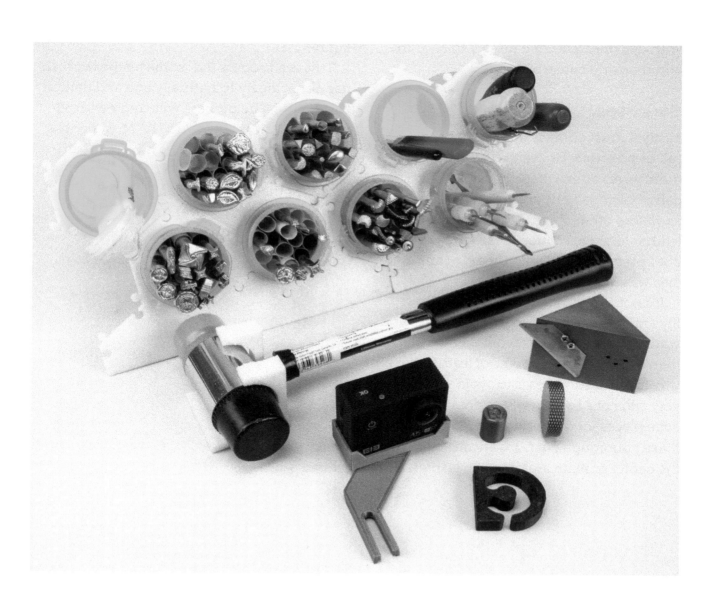

3D PRINTING LEATHERWORKING TOOLS

During the long history of leatherworking, stretching well back into prehistory, the construction of tools for crafting leather has taken many forms. Leatherworkers have used wood, boar bristles, nails, and other materials found close at hand to mold, cut, stamp, and pierce tanned animal hides. The explosion of leatherworking tools in the 20th century expanded ease of use, but the fundamental capabilities would remain basically familiar to a medieval leatherworker.

Using 3D printing techniques to construct leatherworking tools is an obvious extension of the practice of adapting familiar materials and tools to use with leather. The idea of substituting plastic tools for steel may seem blasphemous to some traditionalists, but the convenience and adaptability of 3D printing offers many advantages that make up for the reduced durability. Not all leatherworking tools are great candidates for 3D printed alternatives—plastic shears would have a hard time with 8 oz leather—but there are enough opportunities for use to make this an exciting area of experimentation.

ADDITIVE MANUFACTURING TECHNIQUES

3D printing is a broad term that represents a number of technologies. In general, 3D printing is a form of additive manufacturing. This differs from many traditional manufacturing techniques that start with raw materials and grind, cut, or shave material away, methods that would be referred to as *subtractive manufacturing*. *Additive manufacturing* builds up products from a feedstock of raw material. While plastic is the primary material most people associate with 3D printing, it is by no means the only one used. Ceramics, waxes, metals, polymers, and plastics infused with wood or other materials are used in different techniques.

It's useful to survey the range of additive manufacturing processes. This field is undergoing tremendous flux and innovation. Techniques that may have been unavailable to the hobbyist or Maker last year could be affordable and accessible next year. New processes may enter the market at any time, but most techniques fall within a set of high-level approaches.

PRINTING WITH POWDERS

One of the most common commercial techniques for high-end 3D printing uses very thin layers of superfine powders that are spread in successive layers and selectively hardened; this is referred to as *sintering*. Lasers are the most common method of sintering the material, but electron beams and fusing agents are used in some processes.

Metallic powders allow for the printing of solid metal parts with extraordinary detail. Even high-complexity, heavy-use parts like rocket engine nozzles can be produced in this manner. (See Figure 4-1.)

Powder systems are generally unavailable for home use, and they remain a commercial alternative. But many companies offer these services. Using a hobby 3D printer to prototype and validate a design, then sending it off to a commercial 3D printing company for production in metal or ceramic is a very easy and straightforward approach to making custom tooling or parts.

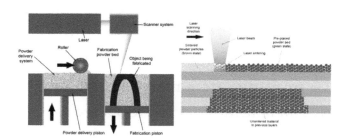

FIGURE 4-1: Selective laser sintering IMAGE COURTESY OF WIKIMEDIA COMMONS

PRINTING WITH LIQUIDS

As the various patents restraining the availability of 3D printing techniques expire, new technologies enter the market. One technology that offers extremely high-resolution printing is broadly referred to as *vat polymerization*. It is similar to the method described for printing with powders, but a thin layer of liquid resin is selectively hardened instead of a layer of powder.

Light, provided by lasers, LEDs, or projectors (typically DLP projectors), is used to harden (aka, *cure*) resin to provide very thin, high-resolution layers. As a result, this is often referred to as *resin printing*. Resin-printed models can be nearly indistinguishable from injection molded products. (See Figure 4-2.)

FIGURE 4-2: A 15 mm long 3DBenchy printed on a formlabs Form 2   IMAGE COURTESY OF WIKIMEDIA COMMONS

While the potential for this approach is incredibly promising, the early-stage offerings tend toward relatively expensive proprietary resin formulations. Build volume can also be an issue compared to other approaches, since the model must fit within the resin container.

Resin specifications vary tremendously. Heat resistance, brittleness, tensile strength, and hardness all depend on the specific formulation. Since most currently available machines use proprietary resins, it is difficult at this time to pick and choose between desired characteristics.

PRINTING WITH PLASTIC FILAMENT

By far the most common approach available to hobbyists and Makers is what is known as fused deposition modeling. This is primarily a process used with spools of plastic filament forced through a heating element and nozzle and laid down under the control of stepper motors. Exciting experimental and early-stage printers using welding wire have shown some promise for metal printing, and paste extruders allow for sugar, clay slip, and other non-filament materials to be printed. But we'll focus the rest of this chapter, and all the 3D printing activity in this book, on plastic filament.

The variety of available filaments is constantly in flux. Different plastics and materials suspended in a plastic binder provide for a huge range of characteristics and capabilities. (See Table 4-1.)

While people constantly experiment and the market is in flux, the vast majority of noncommercial 3D printing occurs with either polylactic acid (PLA) or acrylonitrile butadiene styrene (ABS). Of the two, I tend to do the bulk of my 3D printing with PLA, primarily due to the less offensive fumes.

TABLE 4-1: Filament characteristics

FILAMENT	SPECIAL PROPERTIES	USES	STRENGTH	FLEXIBILITY	DURABILITY	PRINT SKILLS	PRINT TEMP (°C)	BED TEMP (°C)
PLA	Easy to print, biodegradable	Consumer products	★★	★	★★	★	180–230	No
ABS	Durable, impact resistant	Functional parts	★★	★★	★★★	★★	210–250	50–100
PETG (XT, n-vent)	Flexible, durable	All-rounder	★★	★★★	★★★	★★	220–235	No
Nylon	Strong, flexible, durable	All-rounder	★★★	★★★	★★★	★★	220–260	50–100
Flexible TPE, TPU	Extremely flexible, rubber-like	Elastic parts, wearables	★	★★★	★★	★★★	225–235	No
Wood	Wood finish	Home decor	★★	★★	★★	★★	195–220	No
HIPS	Dissolvable, biodegradable	ABS dual extrusion, support structures	★	★★	★★★	★★	210–250	50–100
PVA	Dissolvable, water-soluble, biodegradable, oil-resistant	PLA/ABS dual extrusion, support structures	★★★	★	★★	★	180–230	No
PET (CEP)	Strong, flexible, durable, recyclable	All-rounder	★★★	★★★	★★★	★★	220–250	No
Metal	Metal finish	Jewelry	★★	★	★★★	★★★	195–220	No
Carbon fiber	Rigid, stronger than pure PLA	Functional parts	★★	★	★★★	★★	195–220	No
Lignin (bioFila)	Biodegradable	Looks and feels cool; stronger than PLA	★★	★	★★	★	190–225	55
PC polycarbonate	Strongest, flexible, durable, transparent, heat resistant	Functional parts	★★★	★★★	★★★	★★	270–310	90–105

FILAMENT	SPECIAL PROPERTIES	USES	STRENGTH	FLEXIBILITY	DURABILITY	PRINT SKILLS	PRINT TEMP (°C)	BED TEMP (°C)
Conductive	Conductive	Electronics	★★	★★	★	★	215–230	No
Wax (MOLDLAY)	Melts away	Lost wax casting	★	★	★	★	170–180	No
PETT (tglase)	Strong, flexible, transparent, clear	Functional parts	★★★	★★★	★★★	★★	235–240	No
ASA	Rigid, durable, weather-resistant	Outdoor	★★	★	★★★	★★	240–260	100–120
PP	Flexible, chemical-resistant	Flexible components	★★	★★★	★★	★★★	210–230	120–150
POM, acetal	Strong, rigid, low-friction, resilient	Functional parts	★★★	★	★★	★★★	210–225	130
PMMA, acrylic	Rigid, durable, transparent, clear, impact-resistant	Light diffusers	★★	★	★★★	★★	235–250	100–120
Sandstone (LAYBRICK)	Sandstone finish	Architecture	★	★	★	★★	165–210	No
Glow-in-the-dark	Luminous, fluorescent	For fun	★★	★★	★★	★	215	No
Cleaning filament	Cleaning	Unclogging of nozzles	N/A	N/A	N/A	★	150–260	No
PC/ABS	Rigid, durable, impact-resistant, resilient, heat-deflecting	Functional parts	★★	★	★★★	★★★	260–280	120
Magnetic	Magnetic	For fun	★★	★★	★★	★★★	195–220	No
Color changing	Changes color	For fun	★★	★★	★★	★	215	No
nGen	Like PETG but easier to print, heat-resistant, transparent	All-rounder	★★	★★★	★★★	★★	210–240	60
TPC	Extremely flexible, rubber-like, chemical, heat-, and UV-resistant	Elastic parts, outdoor	★	★★★	★★	★★★	210	60–100

ADDITIVE MANUFACTURING TECHNIQUES

FILAMENT	SPECIAL PROPERTIES	USES	STRENGTH	FLEXIBILITY	DURABILITY	PRINT SKILLS	PRINT TEMP (°C)	BED TEMP (°C)
PORO-LAY	Partially water-soluble	For fun, experimental	★	★★★	★★	★	220–235	No
FPE	Flexible	Flexible parts	★	★★★	★★★	★★	205–250	75

COURTESY OF ALL3DP.COM

THE 3D PRINTING TOOLCHAIN

We discussed the topic of toolchains in general in the last chapter, but the explosion of 3D printing has brought so many new users into the world of digital fabrication that it's worth focusing a bit more attention on the tools used.

The software involved in 3D printing is constantly being improved. The influx of so many users who don't have a prior history in numeric control practices or mesh modeling techniques makes me hopeful that it will force the industry and community to make these tools easier to use and more accessible. The products discussed below are just a snapshot in time of a small selection of what's available and in common use.

To a certain degree, many of the tools involved in digital fabrication suffer from the (evil) programmer's motto, "If it was hard to write, it should be hard to use." Nevertheless, there is a very challenging devil's bargain going on. There are so many highly technical variables involved that developers are forced to choose between ease of use and access to customization. It's not hard to create programs that will successfully print known models on known printers with fixed parameters. It's extremely difficult to offer a novice user hundreds of options, parameters, and choices in an easy-to-understand interface.

As we noted in the discussion of toolchains, the basic activities have to occur somewhere. While integrated tools exist, for the most part, 3D printing tends to break into three classes of software solutions: modeling, slicing, and controlling.

MODELING

It's entirely possible to avoid ever creating your own models. Vast numbers of freely shared models are available at sites like Thingiverse.com and can be downloaded and used with almost any 3D printer. Some of the turnkey 3D printers aimed at kids and specialty uses are intended to work with pre-built models. But I have a strong belief that making and modifying models is the best part of the 3D printing experience.

Models can be created using a wide range of methods. Existing objects can be scanned, CAD programs can allow you to build or sculpt, interactive tools can create variations on existing models by changing parameters, and modeling languages even let you write programs that create objects. No one method addresses every problem perfectly, and the choice of which types you use is a matter of personal preference. I'm going to cover a couple of popular free tools in just enough detail to give you an idea of the tasks involved in modeling.

Autodesk Tinkercad

Autodesk, the company that produces the legendary drafting tool, AutoCAD®, purchased and has been enhancing a free web-based 3D modeling tool called Tinkercad (*www.tinkercad.com*). Tinkercad is designed to look simple, but it is actually a very powerful modeler. Using a concept referred to as *constructive solid geometry (CSG)*, Tinkercad lets users manipulate cubes, cylinders, spheres, and other familiar shapes to construct complex models.

CSG offers powerful techniques; not only can you add shapes, you can subtract them as well. From simple operations, you can build up shapes of essentially unlimited complexity. Tinkercad has the fewest operations and functions of any 3D modeling tool I've worked with, but you can still design outstanding models by grouping shapes that are either solids or holes. (See Figure 4-3.)

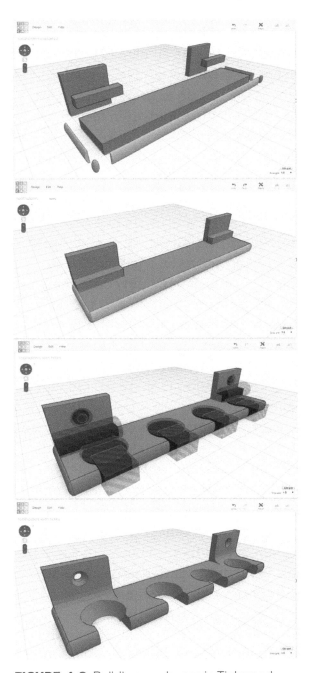

FIGURE 4-3: Building up shapes in Tinkercad

Tinkercad was designed around a very appealing premise. As children, we don't hesitate to make models out of blocks and simple objects. Why should we be more intimidated as adults? Tinkercad provides basic shapes and a few simple operations and leaves the rest up to users. Despite the power it offers, Tinkercad often feels more like play than work.

Autodesk Fusion 360™

Autodesk is doing their best to dominate the 3D CAD landscape, and offers free access for students, hobbyists, and small businesses to their incredibly powerful Fusion 360 web-based tool. Fusion 360 not only offers detailed 3D modeling capabilities, but also provides mechanical simulation to allow the parts to move and interact. The tool includes functionality such as materials stress analysis, 2D drafting, collaboration tools, mesh modeling, and animation.

The learning curve for Fusion 360 can feel steep, but with a few tutorials, you can start making parts fairly quickly. Parameter-driven CSG is deeply integrated and combined with geometric drafting tools. This means that you can build models with the same approach Tinkercad uses, but you can also specify that the dimensions of the various components are based on formulas. Using formula-driven dimensions, if you change one aspect of a model, other parts will change proportionally. You can also use drafting techniques to draw complicated shapes and extrude them into the third dimension. (See Figure 4-4.)

I can't begin to scratch the surface of Fusion 360 in this book, but it's the primary tool I'm using for the models you'll see. If you're primarily interested in 3D modeling for digital fabrication, and debating which tool to invest a learning curve in, it's difficult not to recommend Fusion 360. (See Figure 4-5.)

FIGURE 4-4: Working with models in Autodesk Fusion 360

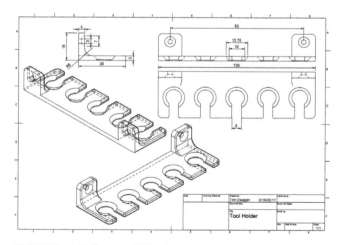

FIGURE 4-5: Fusion 360 drawing output

OpenSCAD

OpenSCAD is a program that leaves many Makers scratching their heads. Rather than manipulating shapes on a screen to create a model, you write code to describe the model. This may sound like an excruciatingly painful way to make 3D objects, but I have to tell you that in many cases, and for many problems, it is wildly easier than graphical modeling tools.

You may still be skeptical, and I should be clear that I am an unabashed OpenSCAD fanboy. It's open source and free and a labor of love from some fantastic developers. While I don't qualify as one of the fantastic developers, I love the program enough that I contributed the code for the 3D scale display.

Many times, when working in a graphical modeling tool, getting dimensions or placement exactly the way you want it can be unbelievably frustrating. This is largely an issue of projecting a 3D image onto a 2D screen, but getting things accurate and precise is something that takes up more time than it should. OpenSCAD gets around this by specifying exactly where and what size things should be. As you make changes, the results are displayed on a 3D screen. You can navigate around the screen, but you can only change the model with (very simple) code. (See Figure 4-6.)

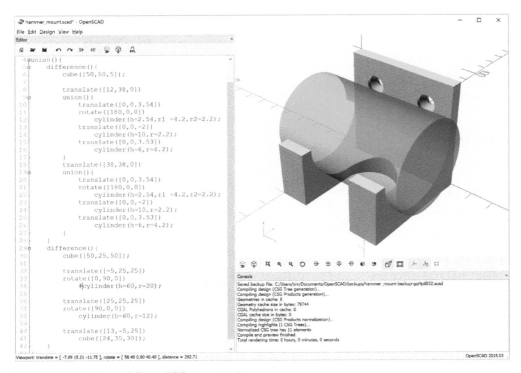

FIGURE 4-6: OpenSCAD CSG description

If you're looking at models in Thingiverse, you may notice that a huge number of them allow you to open them in the Thingiverse Customizer. This is a tool that allows you to change parameters on a model. I have a thumbscrew customizer model at Thingiverse that allows you to select the size, number of bumps, and a wide variety of other parameters to design your own thumbscrew.

Thingiverse can do this because the models used in the Customizer were written in OpenSCAD and used variables for various dimensions or values in the code. The Customizer can display these as sliders, dropdowns, or text boxes that users can interact with. (See Figure 4-7.)

OpenSCAD can, like the other tools, import STL and other 3D model file types. You can then merge, cut, and extend those models to make your own unique changes. (See Figure 4-8.)

Boundary Definition and Mesh Modeling

Having exclusively discussed CSG modeling so far, I'm going to lightly cover another modeling approach. *Boundary definition* is another approach to 3D modeling that defines the shell of an object. This can be done with a variety of approaches. When creating models for video games, movies, and animation, tools such as Rhino or Blender are generally used.

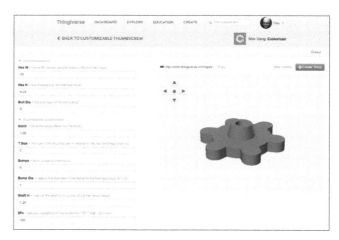

FIGURE 4-7: Thingiverse Customizer

FIGURE 4-8: Working with imported models in OpenSCAD

Mesh modeling is a boundary definition technique that relies on defining the outer surface of an object with polygons, frequently triangles. It's not uncommon for a detailed model to have many hundreds of thousands of triangles in its mesh. (See Figure 4-9.)

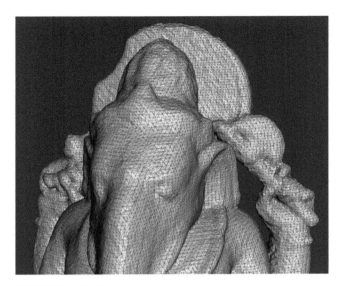

FIGURE 4-9: Mesh model

Mesh models can be created from 3D scanning objects with lasers or photographs (photogrammetry). They can also be constructed with software tools that allow a "sculpting" experience. Tools like Autodesk (yes, again, Autodesk) Meshmixer are powerful mesh editing and creation tools.

Boundary definition and mesh modeling techniques are tremendously powerful, but beyond the scope of this book. They're also used less frequently by 3D printing enthusiasts than the other modeling approaches I've discussed.

SLICING

Once a model has been defined, the toolpath that guides the extruder head has to be created. 3D printing relies on slicing programs to create a toolpath for horizontal layers through the model. Defining the *layer height* tells the slicing program how thick of a slice to take through the model at each layer. The smaller the layer height, the greater the detail. Of course, this also increases the number of layers and the time it takes to print. High-resolution prints at a 0.05 mm layer height look beautiful. (See Figure 4-10.)

Slicers have a complicated task. Not only do they have to determine the boundary of the model at each layer, they also need to determine and react to other aspects of the model so that the full 3D print comes out as desired.

If the layer, or some parts of the layer, are near the top or bottom of the model, a different printing pattern will be used to create a reinforced outer shell. The section of the layer that will be internal to the model gets a special pattern called *infill*. Infill patterning allows the use of less plastic. Rather than

FIGURE 4-10: Very small layer heights

THE 3D PRINTING TOOLCHAIN **53**

printing a solid model, the model can be filled with hexes, squares, lines, or other patterns at a solidity of 10 percent (or any other selected percentage). (See Figure 4-11.)

The slicer also must determine if later layers will be overhangs. This would cause the extruder to squirt plastic into the air, since there is nothing under it to support it. If you choose the option to configure supports, the slicer will create a special breakaway support pattern to provide a scaffold. (See Figure 4-12.)

Factoring different speeds and extrusion feed rates for different parts of the layer (infill, outer shell, inner shell, support, and other patterns), the slicer requires a great deal of computation to create the G-code for the model. On top of all that, the slicer makes attempts to optimize the path to reduce printing time.

There are potentially hundreds of different parameters relating to slicing. This is the area that causes new users the most headaches, since understanding the implications of all those parameters requires a great deal of research and experimenting. Many manufacturers provide standard settings to assist users in getting the best results from their printers, which helps considerably. (See Figure 4-13.)

This only scratches the surface of the technology and capabilities of slicing programs. Temperature management, fan control, and multiple extruders all add complexity and parameters. While slicing programs are

FIGURE 4-11: Inside a print

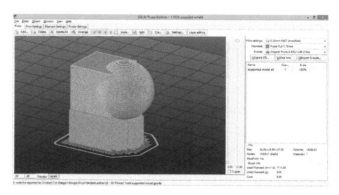

FIGURE 4-12: Printed supports

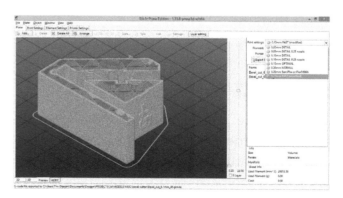

FIGURE 4-13: Prusa slicing presets

typically the most frustrating part of the 3D printing toolchain for new users, advances in interfaces, presets, and eventually machine intelligence will continue to improve these tools.

While Cura and other slicing tools are both powerful and popular, I'll be using the ubiquitous Slic3r program for all the work in this book. It's free and constantly improving. The resources appendix lists more slicers if you want to try different ones.

CONTROLLING

The slicer completes its task by creating G-code for a model. This is the toolpath that must be executed. Once again, there are many choices that I'll list in the resources appendix, but I've always been a Pronterface user and I'm happy with that choice, so that's what I'll be using in this book.

The job of the controller is the most straightforward of any in the toolchain. Feeding the G-code to the printer isn't technically that challenging, but there are other things users want to do at this stage of the process. Many of these, such as the ability to preheat the hot end or jog the tool's position, are really just ad hoc G-code. But one of the main things that has been under development is visualization of the print as it's being printed (See Figure 4-14.)

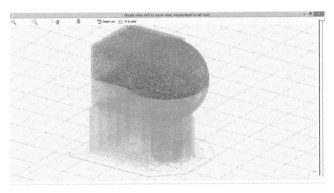

FIGURE 4-14: Controller visualization

The ability to watch the progress of the print reflected on an image of the model usually is more gratifying than essential, but on occasion it can be a lifesaver. On rare occasions, it's possible to save a print that's about to fail in some way, if only to allow you to change to a fresh spool of filament when one is about to run out.

Controllers and slicers are frequently integrated, either by calling one another or, in some programs, by being part of the same application. Most new features that are integrated into a slicer or controller are adopted by the other tools within time. Finding the set that you feel most comfortable with is more important than getting the "best" tool.

3D PRINTED LEATHERWORKING TOOLS

Many predictions have been made about 3D printing's role in a new industrial revolution. While the future remains anyone's best guess, certain realities suggest that many of the most excitable claims are probably hype. Nevertheless, there are some areas in which 3D printing is a tremendous boon to hobbyists and Makers.

With a willingness to trade durability and strength for customization and availability, leatherworkers can easily 3D print many items that expand their capabilities. The expectation that a plastic stamp will withstand 30 years of hammering, as a quality metal stamp would, is unrealistic. As is the idea that the current generations of home 3D printers would produce sharpened metal tooling. But stamps that meet a specific need and can be replaced on demand, even if they are only viable for a couple of projects, are still very worthwhile. The ability to design customized holders that use utility blades to make specific cuts is very exciting, as well. And the capability to add organization to your workplace with customized tool holders is significant to anyone who has stared at a bench piled high in scraps, tools, and materials.

We will explore specific instances of these tools in various project chapters. In this chapter, I'll talk about the high-level issues of printing your own leatherworking tools.

STAMPS

Tooling leather with metal stamps, and stamps in combination with swivel knives or other tooling, is incredibly popular for good reason. Not only are an infinite number of customized results possible, so many stamps exist that anyone who has the collecting urge can get a little crazy in trying to possess as many as possible. (See Figure 4-15.)

As noted in a previous chapter, stamps are typically used by hammering them onto wet leather with a non-metallic hammer. This is a less-successful strategy with 3D printed stamps. It's possible to make stamps that withstand pounding, at least for a while, but it's an easy way to expose their weaknesses.

FIGURE 4-15: A modest collection of leather stamps

Ultimately, the goal is to assert pressure on the stamp against the leather. The hammer or mallet does so in brief, intense raps. But a slower, steady pressure can achieve the same results, sometimes with additional control. I am a fan of using a desktop, 1-ton arbor press as an alternative to hammering stamps. (See Figure 4-16.)

The arbor press uses mechanical advantage to exert pressure against the stamp and leather. It does limit the placement of stamps on large items because of the depth of the throat. The *throat* is the space between the tool head and the stand. I've found that I'm usually able to manipulate the leather so that I can get the press positioned where I want it.

When using a 3D printed object as a stamp, the characteristics of the model can make a difference, as well. The depth of the negative aspects can be surprisingly small and still be a great stamp. If the negative depth is too large, it makes the features unstable and easily broken. Experimentation is essential, but I find that 0.2 mm is usually enough to get a fair image and 1 mm is more than enough. (See Figure 4-17.)

Metal stamps tend to have a thin rod as the handle. This is a weak point in a 3D print and should be avoided. It's also not necessary for the stamp to be very long. It's useful

FIGURE 4-16: A 1-ton arbor press

FIGURE 4-17: Minimal relief depth on a printed stamp

to visualize rubber stamps. They typically have a rectangular body that isn't more than 20–30 mm tall. If you're using an arbor press, then anything larger than the head of the press is simply to provide stability to the stamp relief.

A variety of 3D printing post-processing coatings have become available to Makers. Especially popular with PLA prints, these epoxy coatings add strength, increase hardness, and smooth out striations on the print. I've used XTC-3D from Smooth-On (*www.smooth-on.com*) successfully with printed stamps and generally consider these coatings an improvement. (See Figure 4-18.)

We covered infill earlier in the chapter. Most 3D prints can get by with a surprisingly low level of infill. Commonly, I use 15 percent infill for my prints, but sometimes I use much less. When printing stamps, it's worth spending the time and filament to print something strong. I find that a minimum 60 percent or denser is best, using 80–100 percent for stamps I want max strength from.

MOLDS

Leather cases and holsters are extremely popular projects. Custom-fit molded leather makes a superb material. But few people are enthusiastic about wrapping wet leather around their phone, gun, or other precious object and leaving it there to dry. Traditional responses to this are either to wrap the object to be molded in plastic wrap or to construct a model of the object to use in its place.

If you're creating a model of an object for use in molding leather, 3D printing is an outstanding way to go. Not only are the models waterproof, they can often fill in areas that would be troublesome during the molding process. An example of this might be the trigger guard on a pistol. When molding, this would be an area that could easily get depressed, making the pistol harder than desired to remove from the holster or making the lines of the finished product vary from the desired look.

The challenge is the modeling of the object itself. The model needs to be both true to scale and accurately representative of the desired object. This can mean complicated modeling if you're constructing in a 3D modeling program, or extensive mesh cleanup if you're using a scanning tool. (See Figure 4-19.)

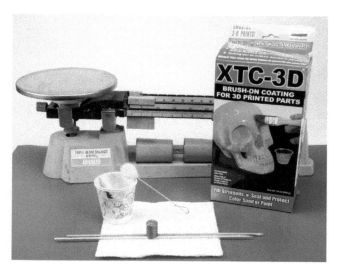

FIGURE 4-18: XTC-3D printed coating

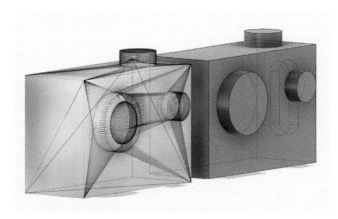

FIGURE 4-19: Mesh vs. CSG model

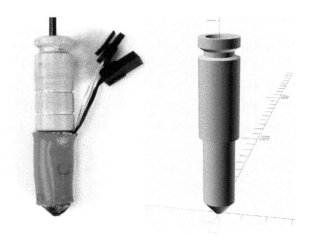

FIGURE 4-20: Original vs. withdrawable shape

Don't underestimate the power of the community. For any number of items, even surprisingly odd ones, there are accurate models freely available. Download if possible; construct if necessary. If you do construct, it's always great to give back by sharing your models with the community.

When constructing, spend some time imagining the model in the leather and skip the effort to model things that aren't necessary to create the mold's shape. It will be tempting to model every detail on the original object, but if the features are on part of the object that won't be inside the leather mold, or are going to be "upstream" of a protrusion that needs to be able to be withdrawn from the mold without snagging, skip it. Better yet, mold the shape that will be pulled out, rather than the shape itself. (See Figure 4-20.)

BLADES

Plastic blades printed on a 3D printer may be useful as scrapers, but they aren't very good as knives. However, if you make plastic mounts, you can insert utility, hobby, or razor blades inside models that serve as guides. Bevel cutting, strip cutting, edge scoring, and a wide variety of other tasks that need blades held in specific positions are great reasons for printing your own tools.

This beveling tool by Thingiverse user Neizam (Shahrom Nurrizam Bin Romli) in Putrajaya, Malaysia, is a lovely design that supports multiple cutting angles. It also allows access to internal nuts without printed supports. (See Figure 4-21.)

When designing 3D printed tools to use with blades, always give some consideration to the mechanism you'll use to extract and replace the blade. It may be tempting to design tools with adjustable blade settings (go for it!), but since you can easily create modifications, consider making multiple tools with different angles or blade placements if it's easier to design.

ORGANIZATIONAL TOOLS

While seemingly mundane, 3D printed tool organizers have turned out to be one of the things that make the biggest difference for me in leatherworking. I've created a range of brackets for hanging my tools, and even a customizable system using pill bottles for storage. Check out *http://www.thingiverse.com/thing:564108*. (See Figure 4-22.)

Take a look around your workspace and think of what tools you want close at hand. Consider which parts of your space are underutilized. Can you hang tools? Is there a good place to mount a hammer on the front of your bench? Do you have a large quantity of some storage device (for me, pill bottles . . . hundreds of pill bottles) that you could create a way to use effectively?

Don't underestimate the time and effort you'll save printing tool holders in your workspace. If you're someone who needs an actual project to get started learning a new skill, creating 3D models to hold tools is a fantastic and easy way to get great results. (See Figure 4-23.)

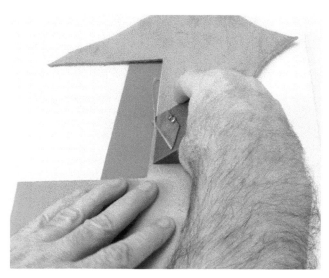

FIGURE 4-21: 3D printed bevel cutter

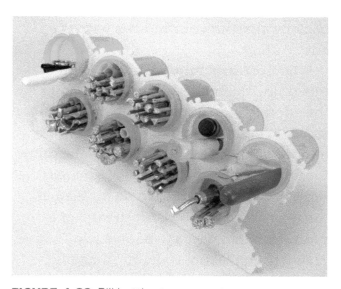

FIGURE 4-22: Pill bottle storage system

FIGURE 4-23: Beveling tool and swivel knife fixtures

THE FRACTAL JOURNAL COVER

I love notebooks. I can't resist buying them in different sizes, whether I have a need for them or not. But of all the notebooks I end up carrying, the Moleskine Cahier is my favorite. These small, simple books are sold as a three pack and have unlined paper that's great for jotting notes, making sketches, or tearing out and leaving with people.

DESIGN

THE BASIC LEATHER PATTERN

The pocket version of the Cahier that I buy is 3½″ × 5½″ with cardboard covers. It has a stitched spine rather than a flat spine, which is important when determining the cover dimensions needed. I wanted to create a simple leather cover for the notebook that would still fit in my pocket, but look more stylish. The Cahier is also easy to bend or damage, so the leather cover will provide additional protection.

Unfolded, the notebook is 7″ × 5½″. We need to add additional margin around it for stitching, plus we need to accommodate the fold when the notebook is closed. Leather is amazingly moldable, but even using 3–4 oz leather, it won't fold as sharply as the cardboard on the notebook cover. We need some additional leather in the pattern at the fold to allow for curvature when the cover bends. One way to determine how much is to fold some dry leather over the notebook and see how much it takes to cover it. (See Figure 5-1.)

We'll use a carved notch down the inside middle of the leather to enhance its ability to make a sharp fold, but it's still best to allow the extra material for a margin. If you've chosen to use a different notebook for this project, use the dry-fold test as a method to determine how much extra leather to allow to cover the fold. If your notebook has a flat spine, you'll need to accommodate two folds. This would mean adding two notches at the appropriate distance rather than the single fold we're using, and you'll still need some extra.

Once you have determined the amount of leather to cover the width of the notebook cover, you'll need to add a ¼″ margin on all four sides to allow room for stitching. In our case the notebook (laid flat) was 7″ × 5½″; the leather needed for the fold allowance was a little less than ½″. Adding margins, this resulted in a leather cover measuring 8″ × 6″.

The notebook will be held in place by sliding the notebook covers into pockets sewn on each side of the inside of the leather cover. The leather for these pockets is the same height as the cover, but will be

FIGURE 5-1: Testing the fold allowance

2" wide. This means we'll also need a pair of 2" × 6" leather strips to use as pockets. (See Figure 5-2.)

THE DIGITAL IMAGERY

I first read Benoit Mandelbrot's book, *The Fractal Geometry of Nature,* in the mid-1980s and immediately became entranced by fractals in all their glory. Fractals come in an immensity of shapes and patterns, but I'm especially delighted by a space-filling curve that Mandelbrot published in 1982 that is known as the Mandeltree. (See Figure 5-3.)

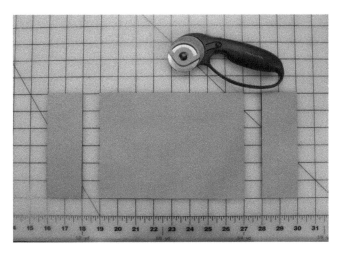

FIGURE 5-2: Leather pieces for the journal cover

The Mandeltree evokes the beautiful meditation labyrinth at Chartres Cathedral, a single line that weaves in and out and around to form a complex image. (See Figure 5-4.)

FIGURE 5-3: Mandeltree IMAGE COURTESY OF WIKIMEDIA COMMONS

FIGURE 5-4: Chartres Cathedral labyrinth IMAGE COURTESY OF WIKIMEDIA COMMONS

For the inside pockets of the journal cover, the classic Julia set supports our fractal motif. (See Figure 5-5.)

Below we have the completed pattern for our project. (See Figure 5-6.)

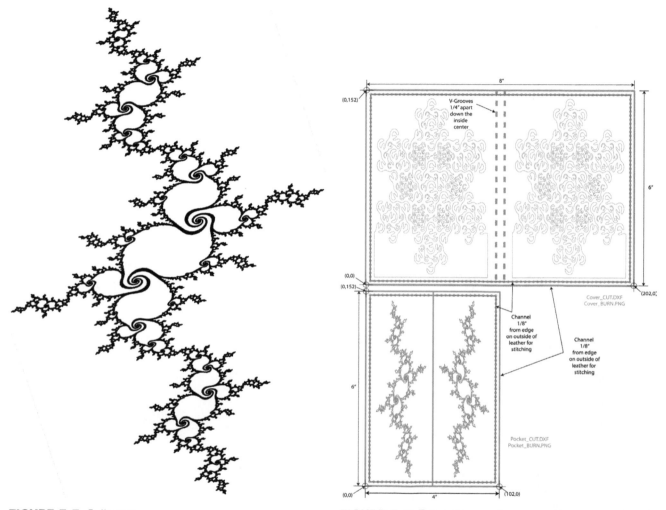

FIGURE 5-5: Julia set

FIGURE 5-6: Fractal journal cover pattern

DIGITAL LASER FABRICATION

We are starting with bitmapped (or *raster*) images of the fractals we want to burn into the leather. For this project, we'll use the tracing function of our software to create vector outlines of our images. These will burn faster and look crisper than raster images. The choice of raster or vector relies on the needs of the imagery and project, so it's hard to make a general rule about whether one approach is better than the other. In the next chapter, we'll explore the details of burning raster images.

PREPARING THE FILES

The first action will be to cut out the leather using the laser. The ability to cut is directly related to the power of the laser. I'm using a 2500 mW blue-violet laser, which is generally considered to be the weakest laser with any real cutting ability (at least with leather). The ability to make an effective cut also depends on the thickness of the leather. This project is using 3–4 oz leather, which can be cut in a single pass on the 2500 mW laser. Heavier leather can still be cut, but may take additional passes.

The two primary variables when working with the laser are speed (measured in mm/min) and power. Since we're burning leather, we risk excessive charring if we use too much power, or if we move the laser too slowly for the power level chosen. Cutting on an engraving laser generally relies on max power, which is typically described as level 255 (out of a range from 0–255). Even so, we have to slow the laser down considerably. For this thickness, a feed rate of 45 mm/min is slow enough to make the full cut in one pass without charring the leather too much.

It's a perfectly valid choice to speed the laser up, thereby reducing the charring, and make the cut in multiple passes. Speaking from experience, I'd rather succeed in one cut, if viable. Every time you set up a pass on any digital fab tool, you risk a mid-process fail. These fails can occur due to hardware or software glitches. Over time, they happen far more than any hardware or software manufacturer will ever admit. It can be monumentally frustrating to have a piece of work ruined on the last pass of a multi-step operation. But it happens.

In this project, we're not only cutting out the leather pieces, we're also going to use the laser to produce the holes for stitching that are usually pierced with an awl or punched with a special fork. (See Figure 5-7.)

FIGURE 5-7: Manual stitching-hole creation

This requires a vector file that has the desired outline of the leather piece and circles for each hole. I've made two separate files, one for the outside cover and one (to be cut twice) for the inside pockets. These could be combined into a single file, but I wanted to reduce the impact of a fail, and make smaller cuts.

I produced the files in Fusion 360 and exported them as DXF files. Luckily, the T2Laser software I'm using has a DXF Optimize function. Even though I went to great lengths to try to place the circles in careful order, the normal conversion to G-code resulted in a common problem where the order in which the objects in the DXF are undertaken is, for all intents and purposes, random.

To see what I'm talking about, you can see the toolpath that results from the conversion of a very carefully constructed DXF for the journal cover. (See Figure 5-8.)

This meant that the laser was constantly making movements back and forth as it jumped around from point to point. This takes longer and increases the opportunities for the machine to lose registration. A clean, well-ordered toolpath is a balm to the soul. Using the DXF Optimize function cleaned things up so that the toolpath looked a lot better. (See Figure 5-9.)

As a note, it is an *extremely* good idea to use a tool to visualize your G-code toolpath prior to operation. Nicolas Raynaud has posted a very handy online G-code simulator at *https://nraynaud.github.io/webgcode/* that will let you visualize a toolpath without having to sit through a pass on cardboard or, worse, ruin a piece of leather.

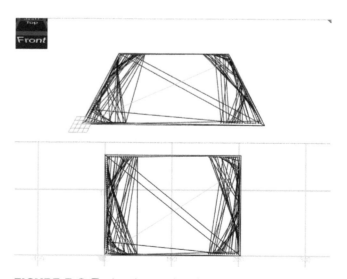

FIGURE 5-8: Toolpath spaghetti

We will need four files for this project: Cover_Cut, Cover_Burn, Pocket_Cut, and Pocket_Burn. The DXF _Cut files will provide vector images for cutting out the workpiece shape and stitching holes. The controller software will optimize and convert these to G-code. The PNG _Burn files will provide images from which to engrave the fractals. The controller software will auto-trace these (to turn them from raster into vector images) and convert them to G-code.

I will be primarily using Autodesk Fusion 360 to create the DXF files and Adobe Illustrator for the PNG files. There are many different programs that you could choose from for your work. Fusion 360 is free for hobbyists and students and well worth the learning curve. Inkscape and GIMP are also excellent choices.

For the Cover_Cut file, we need to create a rectangle with the dimensions 8″ × 6″. We'll add 1⁄16″ holes every 1⁄8″ of an inch all the way around, centered 1⁄8″ from the rim. (See Figure 5-10.)

For the Cover_Burn file, we need an image file with 8″ × 6″ dimensions and our Mandel-trees placed where we want them. Use a white background and convert the Mandel-tree image to black and white. Position the images so that they are evenly spaced from the top and bottom of the image boundaries. This spacing should also be used for the distance each tree is from its respective side. (See Figure 5-11.)

The Pocket_Cut file is a pair of 2″ × 6″ rectangles with a matching hole pattern to the cover. (See Figure 5-12.)

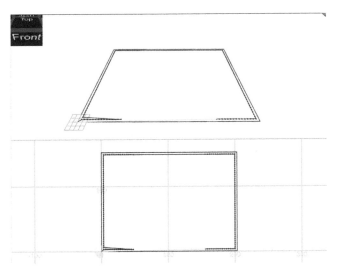

FIGURE 5-9: Optimized toolpath

FIGURE 5-10: Main cover cut plan

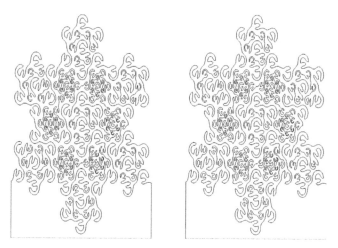

FIGURE 5-11: Main cover burn plan

FIGURE 5-12: Pocket cut plan

Handling Dimensions

Handling the dimension changes that result from moving files between programs is unbelievably frustrating. I offer the following advice: worry about aspect ratio more than actual dimensions. If you work with high-resolution images that have the aspect resolution you ultimately desire, you can resize to the desired dimensions in the laser software during the final step. Trying to get all your programs to agree on dimensions will prematurely age you. DXF files don't even define what units they are in, so you can get switched from mm to pixels to points to inches as you move things around.

The Pocket_Burn file is a 4" × 6" white background with black Julia sets positioned to match their locations on the cut pieces. I stretched the Julia set in the long axis to better fill the space. (See Figure 5-13.)

FIGURE 5-13: Pocket burn plan

SETTING UP THE MACHINE

The budget laser tool I'm using does not have limit switches. This means that you cannot *home* the machine. Homing typically involves commanding the machine to travel in a negative direction until it hits a limit switch, then stopping. The location where it stops is then defined as zero on the axes chosen for homing. Without limit switches, the machine cannot find zero by itself.

It might be tempting to simply run the machine to the end of its travel and declare that the origin. The problem with this is twofold. The first problem is that it's abusive to the machine. The stepper controller doesn't know it's at the end of travel; it continues to grind away at the belt or screw until you decide to stop torturing it. This isn't good for the components and possibly introduces slippage or other problems.

The second issue is potential inconsistency. Using a mechanical end point means that there may be variance on where things stop based on the speed or torque at which the end of travel was encountered. There may be bounce-back. This might seem trivial, but having a few tenths of a millimeter difference in your origin from homing to homing should be unacceptable.

G-code

Over the years, there have been a number of languages created to communicate with digital fabrication equipment. The most common of these is G-code (RS-274). Dating from the early 1960s, G-code has been used in all kinds of *numeric control* (the *NC* of *CNC*) systems. There are a vast number of variations of G-code, and many versions implement the same commands in different ways or add functionality to the language.

G-code is typically intended as a bridge language, generated by one system and read by another, without humans necessarily paying any attention to it. Most implementations lack loops, conditionals, and programmer-defined variables. Most instructions consist of a letter to indicate the type of command, followed by numbers or additional letters.

The majority of the commands are *G* commands, indicating the type of motion, or *M* commands, indicating a miscellaneous function. Over the years, every letter of the alphabet has been used in one way or the other, but G and M codes compose the bulk of commands in most programs that home users will encounter.

Most of the time, the G-code is executed without being examined, but on occasion, hand-editing G-code, or even directly controlling a machine via G-code commands, can be very useful. To understand the specifics of the G and M codes relevant to the machine you're working on, you have to find a reference; even different versions of the same implementation may use commands in different ways.

A tiny example of a simple RepRap G-code program might look like the following:

```
G21                    ' set units to mm
M190 S180              ' wait until bed temperature is 180°C
G92 Z0                 ' set Z axis value to Zero
G28 X Y                ' home X and Y axis (move to endstop positions)
G1 F1250               ' set the feedrate to 1250mm/min.
G1 X10.0 Y10.0 F1500   ' linear move in X & Y axes to (10,10), accelerate to 1500mm/min.
G1 Y100.0              ' linear move in Y axis only, position now (10,100)
G1 X100.0              ' linear move in X axis only, position now (100,100)
G1 X50.0 Y50.0         ' linear move in X & Y axes to (50,50)
G28                    ' home all axes
M140 S0                ' set bed temperature to 0°C (turn off the heater)
```

Finding a reference that explains the G-code implementation for your device is critical. Search the web with the term *G-code* and the name of your machine. It is especially useful to read forum posts relevant to your machine for details about specific G-code commands.

It is possible to use calipers to manually position the head to a predetermined home. This is worth doing on occasion if only to make sure that the *x-gantry* (the frame component that the tool rides on to move in the X dimension) is truly perpendicular to the system frame. Find locations to position the calipers so they are not at an angle, and use a crisp visual mark on the machine as the point to measure against. In my case, I use the acrylic frame of the *x-carriage* (the component that holds the tool to x-gantry) and the end of the beam it is mounted on. (See Figure 5-14.)

This is a tedious operation to undertake for every cut or burn, and it's overkill unless there is some reason you need to orient the workpiece the same distance from the frame each time. Usually you want to have a quicker method for using the machine and it's not critical that the origin be in the exact location relative to the frame.

The solution is to pick an arbitrary location in the work area and declare it the origin. All software packages should provide a command that does this. (See Figure 5-15.)

Once the origin has been set, I want to create a layout guide on the bed so I can position the workpiece. On the bed of the laser engraver, I've placed a piece of light-colored paper covered by a sheet of wax paper. The light paper serves as a fresh "target" surface, and the wax paper helps protect the bed from the damp leather (I'll talk about why the leather is wet a little later

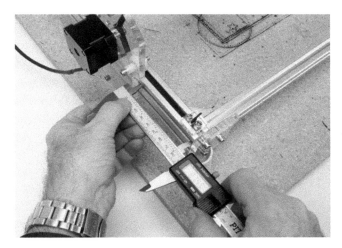

FIGURE 5-14: Manual position setting

in the chapter). With the laser head positioned in the location I've chosen as my origin, I pulse the laser to create a burned point on my paper target. (See Figure 5-16.)

In the laser control screen of the software I'm using, I use the jog controls to move

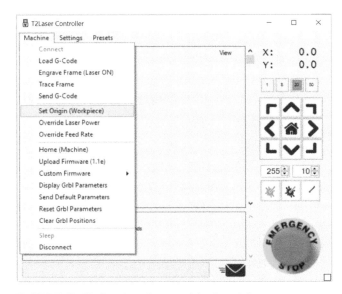

FIGURE 5-15: Setting the machine's origin

the laser 50 mm in the positive X direction and pulse again. I continue to do this up through 200 mm, and then I return to the origin and do the same thing on the Y axis. (See Figure 5-17.)

I could have burned lines directly onto my bed, but I haven't permanently attached the laser tool to the waste board it's standing on.

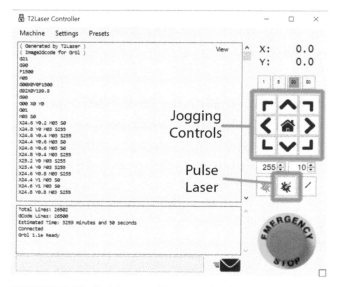

FIGURE 5-16: Pulsing and jogging the laser

FIGURE 5-17: Creating the layout guide

This is partly due to the difficulty in attaching the acrylic legs of the tool, and partly because I want to easily move the laser to different surfaces. I pay for this by having to create the layout guide fresh each time the laser engraver is moved. In fact, I do it a lot more often than that, since the laser, when cutting, marks up the layout target, and I generally want a fresh one for each new workpiece even if the laser engraver remains in place.

Return the laser to the origin position when the engraving operation is completed. Most laser engraver software packages will set the position of the laser to (0,0) when you open the control screens. They do this because there is typically no method to determine where the laser is positioned. If we had limit switches, we could use them to return to a known position, but without those sensors, we'll have to park the head in the location we have defined as (0,0) prior to any start of the control software.

PREPPING THE WORKPIECE

In this project we will be using the laser to cut out the workpieces. Arguably, you could place the laser engraver on top of the entire section of leather you're using (single shoulder, double shoulder, whatever), but it's far more effective if the piece you're cutting from fits inside the tool's work space. I prefer to rough-cut the leather slightly larger than the desired piece, and then use the laser to cut the final piece out of that section. Since we're ending up with one large and two smaller pieces for this project, I rough cut two pieces of leather. The first is about 8½" × 6½" for the 8" × 6" main cover, and the second about 5" × 6½" for the two 2" × 6" interior pockets. You don't have to be careful to make clean or perfectly even cuts for these. (See Figure 5-18.)

Leather can be worked on the laser tool in a wet or dry state. The characteristics of how both of these work is different. In general, I prefer doing all the work wet. This reduces the smell, smoke, and charring that occurs. Wetting the leather is easy; just hold it under running water so that it gets evenly wet on both sides. Do this for 30–45 seconds. This is sometimes referred to as *casing* leather, and is necessary to mold or tool the workpiece.

Casing for tooling is a little different than getting leather wet for laser work. When tooling, you don't want the leather soaking wet, just evenly damp. Regardless of how wet you want your leather; the primary concern is that is it wetted evenly. Dry or unevenly moist spots will cause problems when the leather is drying out and when you try to apply dye or sealants. Many leatherworkers will leave their wetted leather sealed in a plastic bag for hours to soak it evenly.

For laser work, we just need it evenly wet. Since we're working with relatively thin 4–5 oz leather, it isn't difficult to get it soaked through. We're not tooling or molding, so we don't have to be as careful with this process. It's not critical how wet the leather is; anything less than dripping will work. (See Figure 5-19.)

We'll cut the interior pockets first, then the main cover. We'll perform the cutting operation followed by the burning operation without moving the leather. That way, we don't have to reregister the piece. Using the layout

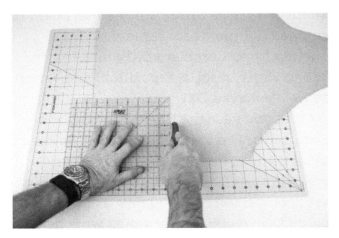

FIGURE 5-18: Making the rough cuts

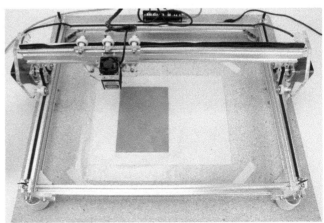

FIGURE 5-19: The pocket piece, wet and ready to cut

guides burned into the paper and placed on the bed (perhaps taped to keep it from moving), place the leather so that it covers the dimensions of the piece to be cut.

Next we'll check the focus of the laser on the leather. The focus point of the laser should be as crisp as possible on the top of the leather. The laser control board should provide a low power Laser On button. Turn this on and focus the laser by turning the collimator that is attached to the laser head. I like to use a magnifying glass to help me be sure the point is focused. (See Figure 5-20.)

READY TO CUT

Import the Pocket_Cut.DXF file into the laser software. Check the dimensions, and

> ### Consistent Cutting
>
> The layout guides indicate where the edge is going to be cut, so you want to position your rough-cut piece so that it overlaps these guides. The most consistent cut results from consistent conditions. If you're trying to cut right up against the rough cut you've made, rather than through leather on both sides, the heating and resulting cutting will be different on that edge. Position the leather so that the laser will always have leather on both sides of the cut.

change them if necessary to the desired dimensions. It's entirely possible that they will be outrageous or negative numbers and will not resize. This is usually the result of a DXF exported from a program that is using very different units, or has an inverted grid. I have, for the most part, stopped fighting these issues and frequently import problem DXFs into Fusion 360 to verify, rescale, reposition, and export as well-behaved DXF files.

Once you have the DXF in the laser software and scaled to the desired size, we need to set the feed rate and laser power levels. For cutting 4–5 oz, wet leather, 45 mm/min at full power (for my 2500 mW laser) works pretty well. A little slower would work better, but the circles would char more. Any place

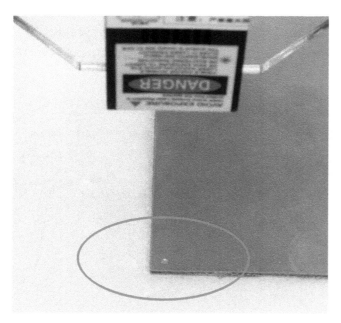

FIGURE 5-20: Focusing the laser

where the laser path gets too close to itself brings with it the risk of overheating the material.

Since we're cutting using a vector file, we don't have to concern ourselves with some of the other settings involved in raster work. These include settings to move the laser horizontally or diagonally, use PWM grayscale, use dithering, or other options. We'll explore these in a later project.

Leather Shrinkage

Unfortunately, there is a problem with sizing leather when it is wet. It will expand, frequently in only one direction. A piece that measures 6" when wet may shrink to 5¹³⁄₁₆" once dry. The type of shrinkage will vary by the weight and quality of the leather and the cut of the piece in relation to the grain of the hide.

There are three ways to deal with this. The first is to cut a test piece from the leather section you'll be using that is in the same orientation as the piece you'll cut for your project. Cut it to a known length and width and measure carefully. Wet it fully and measure to determine the direction and scale of expansion. Compare this to the original measurements to determine expansion factors and increase the size of your pattern to account for it.

I'll be honest, that's a lot of work and must be done for every new section of leather you buy. Even if you're very careful, it's difficult to be extremely precise in the results. Leather is grown, not manufactured, and has inconsistencies. That's frequently part of the charm of a leather product. It feels more organic.

The second method to deal with the problem is simply to oversize your pattern by about 4–5 percent so that you have room for shrinkage. The shrinkage in my project was 3.125 percent. That's ³⁄₁₆" and really didn't make a world of difference. As long as the pieces don't shrink so much they don't fit anymore, it'll all work out. And if they do shrink, you can frequently wet the leather and stretch it to make it work. This is my preferred way to deal with shrinkage.

The third method is to cut the leather dry. This is perfectly viable, but it does require that you use lower power, a faster feed rate, and a multi-pass cut to avoid charring, especially in the stitching holes. It also results in a lot more smoke and stink. There are times when this is the best approach, but it takes a number of test cuts to get the parameters working well.

> ### *Run a Cardboard Check*
>
> Before you sacrifice your leather, or if you're having trouble getting things to work right, it's best to substitute cardboard for leather for a couple of attempts until you're seeing the results you want. You'll need to change the feed rate, speeding it up so that you don't over-burn the cardboard. Cardboard is cheaper than leather, so I highly recommend this step when you're getting used to your laser.

At this point, with the image sized appropriately and the feed and power set, save the G-code file. The G-code in T2Laser is saved as an NC file, which is a human-readable text file. I have developed the habit of adding significant parameters to the NC file name so that I can distinguish between different experiments on the same source file. In this case, I named the file Pocket_Cut_45-255.NC to indicate the 45 mm/min feed rate and power level of 255.

Click the appropriate button to send the G-code file to the laser control screen for execution. (See Figure 5-21.)

It can be very useful to have the software trace the outline of the frame that encloses the outer edges of where the laser will travel. Most packages will allow you to do this with the laser on or off. Choosing off, we can get a rough idea if the laser is likely to travel outside the leather we want to cut.

There's not much else but to command the laser to go forth and scorch. All packages should have an Emergency Stop button that you hit if you need to immediately shut the laser and motors down. Many packages also have a Pause function that you can use to stop the motors and turn off the laser, but resume again later. Be careful if you use Pause not to bump or move anything; the precision that the laser is capable of is also

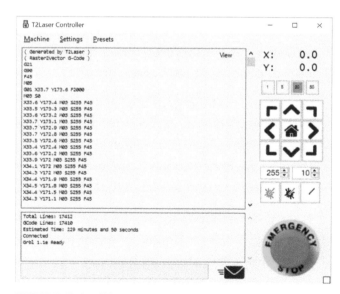

FIGURE 5-21: File ready for cutting

painfully evident when things get jarred or moved. (See Figure 5-22.)

FIGURE 5-22: Leather that was moved during burning

Fume Safety

When cutting wet leather, there is still smoke. Some of it is steam, and there is less of it, but it is burning leather nonetheless. Burning leather does not smell as appealing as cooking meat, but the smoke is generally not a health hazard in the small amounts produced. Nevertheless, if you're using a laser to burn things, you should prepare some form of fume extraction. This might be as simple as a fan in the window, or it might be a more elaborate setup. If you're using the laser on any non-organic substances, especially acrylics, plastics, or other petrochemical products, it is *essential* to remove the fumes. Do not risk your health, or anyone else's!

READY TO ENGRAVE

Since we have the leather workpiece right where we want it, we'll leave it where it is and engrave the pattern onto it. This ensures we will have the best registration we can achieve. If you're not doing something that requires high precision, you can reposition the leather, but it's a lot easier not to.

Exit from the laser controller and import the graphic file Pocket_Burn.PNG. Check the dimensions to be sure they match the target, and rescale if necessary. Set the power to 255 and the feed rate to 1000 mm/min. The feed rate is higher so that the laser will burn, but not burn through, the leather. There's a fairly wide range of feed rates that will work, so experiment on scrap leather to determine which works best for you.

We could burn the image onto the leather as a raster bitmap or a vector. I prefer the vector approach whenever feasible. The raster images can look great for actual bitmaps, but for line art, vectors really shine. The Julia set image we're using here could go either way, but we'll choose vector in this project.

Select Auto Trace . . . from the Edit menu. Some variation on this command should be available in your package. If not, you can achieve a similar result using Inkscape, Illustrator, or some other vector image–editing program and exporting to DXF. T2Laser's Auto Trace does a surprisingly good job on most images. (See Figure 5-23.)

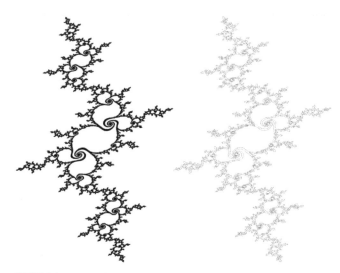

FIGURE 5-23: Original and traced Julia set

Save the G-code file. Based on my naming standard, I'd name it Pocket_Burn_1000-255.NC. Send it to the laser control screen and burn it onto the pair of pieces we previously cut.

CLEAN UP AND TRIM

When the operation is completed, you can remove the workpiece from the bed. You may find that the laser didn't completely cut through in a few locations. This can be remedied in future passes by reducing the feed rate, but at the cost of additional charring. If there are only a few spots uncut, I use a hobby knife or razor blade to carefully cut through and free up the piece.

Holes punched in leather shrink over time. They shrink worse if they were cut when the leather was wet since the entire piece will shrink as it dries. The laser has cauterized the holes' edges, so they won't shrink as badly as a hole punched with a fork or an awl, but it is still important to clear out any charred leather bits or other detritus while the hole is still at its maximum. Once the leather has completely dried out, use a leather-stitching needle or similar instrument to clear all the holes. (See Figure 5-24.)

FIGURE 5-24: Clearing the stitching holes

Avoid Carbon Stains

Before handling the piece very much, it's worth running it under some water for a moment or two. The laser has burned the leather and left char (i.e., carbon) that can get rubbed into the leather and stain it if you're not careful. Rinsing the carbon off the piece will distinctly improve the appearance. If you're using the laser on dry leather, you can use some wide masking tape. Cover the dark areas with tape and pull up the char. You may need to do this a couple of times; you'll be surprised at how much keeps coming off.

CUT AND ENGRAVE THE OTHER PIECES

After the pocket pieces are cut and engraved, perform the full set of operations with the larger leather piece using the Cover_Cut.DXF and Cover_Burn.PNG files for the cover.

LEATHERWORKING

The traditional leatherworking steps that remain are simple. We'll dye the leather, put on a finish coat, and sew the pieces together. This doesn't mean that you couldn't do tooling or other techniques, but they'd be enhancements over the basic project.

CUTTING CHANNELS

We want the cover to fold nicely around the spine of the journal. While leather can be formed to a fold while wet, to get a sharp fold it works best to cut a groove. This channel should be about one third the thickness of the leather and is cut on the rough side of the leather. The spine of the journal I'm using is thick enough that I decided to cut two grooves about ¼" apart down the center. This allowed the leather to make a very even fold.

To make this channel, first use a needle or scribe to mark a line down the center of the cover piece. Then, with a ruler as a guide, use a V-gouge to make the cut. (See Figure 5-25.)

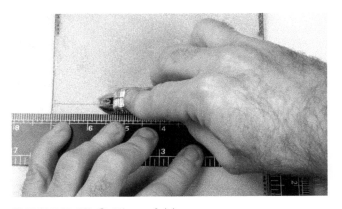

FIGURE 5-25: Cutting a fold groove

There are different styles of V-gouges; the most common are adjustable. It's always important to try them out on a piece of scrap leather before cutting an important piece. (See Figure 5-26.)

Another shallow cut we'll make is on the top side of the leather pieces. This will be a groove that runs along the stitch holes for the thread to sit in. The thread is better protected from wear when it doesn't sit on top of the leather, so the groove helps a great deal over time. Usually, the channel is made prior to stamping or punching the stitching holes, but in our case, the laser made them for us, so we'll do it afterward.

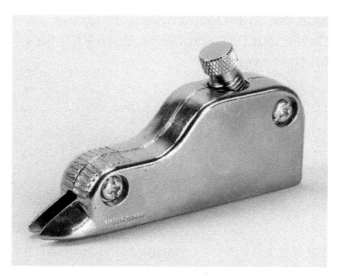

FIGURE 5-26: Adjustable V-gouge

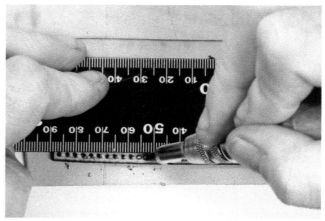

FIGURE 5-27: Cutting the stitching groove

Most stitching groover tools have the ability to set a guide so that you can run the tool evenly along the edge of the leather. Since we want to follow a set of holes that might be a little out of true due to uneven leather drying, it's better to remove the guide and just run the groover along the holes. Place a ruler on the inside to act as a guide rail and to protect the main area of the piece from accidental grooving. (See Figure 5-27.)

COLOR AND FINISH

Dyeing leather can be as basic as brushing dye onto the dried leather or as elaborate as a custom airbrush project. There are a variety of different dyes available on the market and a wide range of colors. I prefer to use water-based dyes, which have come to dominate the market in the last decade. Many leatherworkers prefer the older alcohol- or oil-based dyes, but the lower volatile organic compound (VOC) content of the water-based dyes has been a big factor in their growing acceptance.

There are also leather stains and leather paints. Stains are primarily effective at changing the leather tone, rather than the color. Some manufacturers will combine colors with stains, so the definition isn't fixed. Paints generally sit on top of the leather rather than soaking in like dyes and stains. Most leather paints on the market are acrylic and can be thinned to achieve a range of effects.

There are also antiquing gels. These products are usually wiped on and then wiped off again, leaving the gel in cuts and crevices to achieve a darker effect. We'll be using antiquing gels in some of the CNC-based chapters to highlight the cuts.

Leather finishes are used to seal and protect the leather after dyeing. Once the finish is on, which generally confers a degree of water resistance, the leather won't accept dye, stain, or gel. This usually means that the finish is the final step, but there are circumstances where a finish product is used as a selective resist to create elaborate dye patterns. Finishes leave a satin or gloss result, depending on the product. (See Figure 5-28.)

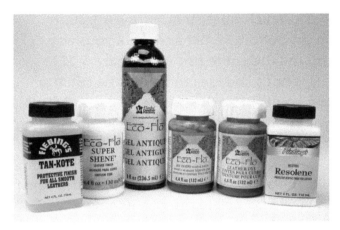

FIGURE 5-28: Leather coloring and finish products

Dye can be manually applied with a brush or a sponge. Care should be taken to evenly apply the dye without creating streaks. Thinning the dye and applying multiple times (drying thoroughly in between) is one way to avoid this. Using a circular motion with overlapping strokes is another.

I have become addicted to the results I get from using an airbrush to apply the dye (and finish). While I own an expensive high-quality airbrush, I've found that, for leatherwork, I prefer using a super-low-budget airbrush from the local import tool store. The reason is multifold. First, the cheap airbrush uses a Venturi effect, blowing air across the top of an orifice on the bottle to spray the liquid. This means that the airbrush itself doesn't require cleanup, since no paint, dye, or finish is passing through it. This design is shunned by most artists because it has very limited control, but it works great for leather dye.

Second, this airbrush uses "quick change" bottles. I can quickly change between dye and finish, or different colors of dye, without having to clean out the opening. I store a range of my favorite products in the bottles, sealing the holes, and find that they're ready for use on a moment's notice.

Lastly, the type of airbrushing I'm doing is applying a broad, even coat. I'm not trying to do fine-line work or complicated fades. If a project needed that, I'd return to my high-end Badger airbrush. But in the last 5–6 years, the low-end airbrush (usually less than $10 USD) has met all my leatherworking needs. (See Figure 5-29.)

FIGURE 5-29: Cheapo airbrush kit

Airbrushing dye (and watered-down finishes) leaves a smooth, even coat without oversaturation, streaks, or blotches. I also use a small desktop airbrush hood to capture and filter the spray. Airbrushes are designed to create lots of small particulate spray and you don't want to breathe it. It's also surprising how much of what you're spraying ends up in a thin layer on everything in the room unless you use a hood. Hoods like the one I use have a filter, but also come with a tube to route the exhaust out a nearby window. Using a hood makes airbrushing practical in areas that would otherwise be impossible. (See Figure 5-30.)

Once the dye is fully dry, you can remove the dye residue and help create shine by buffing with a clean towel, old T-shirt, shoe shine brush, or some sheep's wool. Buffing before finishing is an important technique for getting great results.

STITCHING

The cover has stitching holes around the entire rim, but the pockets only have them on three sides. This is so that, when the pockets are stitched to the inside of the cover, you can slide the notebook cover into the pockets.

The stitching is straightforward: start in the bottom center and work your way around the entire rim. (See Figure 5-31.)

It's not a bad idea to double the stitch at the point where the pocket pieces end. This will strengthen them so that repeated pulling doesn't break the thread at that location.

There are two reasons. The first is that I like the aesthetics of the outside of the cover having a complete rim of stitching. If you decide that this isn't as appealing, you can use a DXF editor (such as Fusion 360) to remove the holes. Here's how it looks with the stitch holes only present where they're needed. (See Figure 5-32.)

The second reason is that, over time, a leather border that has stitching has less

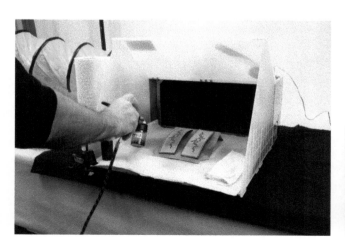

FIGURE 5-30: Small airbrush hood

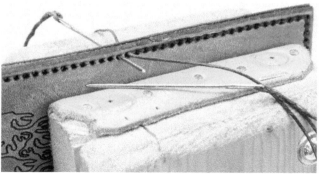

FIGURE 5-31: Stitching the cover

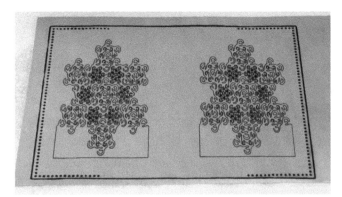

FIGURE 5-32: Alternate stitch pattern

stretching. It's also stronger and better able to handle extended wear. The stitching serves to "ruggedize" the edge of the leather. Over the years, I've started stitching any edges that I believe might be bent or stretched.

THE RESULTS

The journal cover has a wonderful feel. It fits well in my pocket and is going to be wonderful once it breaks in and softens up with wear. (See Figure 5-33.)

ENHANCEMENTS

ACCELERATING THE BREAK-IN

While veg tan leather will soften by itself to a degree with wear and use, it's also possible to soften leather using various products. It's important to choose carefully, since many products that might seem to improve leather will actually contribute to its demise. Recall that leather is really a set of compressed fibers. If those fibers separate, the leather falls apart. Many types of oils will lubricate the fibers in a way that accelerates this process. Any type of oil that oversaturates the leather will contribute to the leather degrading faster.

There are a few types of lubricants that aren't as bad for leather. Neatsfoot oil is a natural oil product originally derived from the shin bones and feet of cattle. While it can contribute to oxidation over time, one or two light coats before putting a finish on leather can help soften it. Olive oil is also used by many leatherworkers (cold-pressed extra virgin has the best chance of not going

FIGURE 5-33: Finished journal cover

rancid over time). Be careful; you don't want to oversaturate the leather. Best practice is to give each light coat of oil several hours to evenly soak in before the next coat.

Other commercial products that are good for leather upkeep can be used to soften, as well. Lexol, Dr. Jackson's Leather Conditioner, Fiebing's Golden Mink Oil, Natur-Glo, and other similar products have all developed strong followings over the years. (See Figure 5-34.)

It's important to come to terms with the reality that veg tan leather just isn't going to be as soft and pliable as chrome tan or latigo leather. The tradeoff is that veg tan is workable and dye-able. Trying to make veg tan ultra-soft and buttery is a great way to shorten the life of the leather.

CUTTING CORNERS

I'm not referring to economizing. I actually mean trimming the corners of the cover and pocket pieces. The cut pattern could just as easily have been a curve or diagonal as a sharp 90° turn. You can edit the DXF files to change this; just be sure to make the corners of the pocket pieces match the cover corners. (See Figure 5-35.)

This may also change the pattern of stitching holes. There is a lot of leeway in how these holes are laid out. You could increase the distance between them, have them follow a curve around the corner, or even put in a double row of them. As long as you leave room for the journal cover to fit,

FIGURE 5-34: My favorite leather conditioners

FIGURE 5-35: Trimmed corners on the cover DXF

you can consider the stitching and things like the corners creative elements as well as structural ones.

CLOSURE

The interior channels we cut should keep the journal reasonably closed, but you might want a way to force closure. There are dozens of ways to add this. You could sew a tab on one side that wraps around and uses a snap closure on the other side. Button studs and magnet clasps are easy alternatives to a snap. (See Figure 5-36.)

FIGURE 5-36: Snaps and studs

Another approach would be to extend the length of the cover panel so it can wrap around the opening and close on the other side. (See Figure 5-37.)

We'll explore setting snaps in the next chapter, "The Ghouls and Gears Multi-Tool Holder," but they're worth considering as an upgrade to this project. The important thing to keep in mind is that adding things like a closure strap may require performing a specific order of operations. A tab closure would need to be sewn to the cover *before* sewing the pocket piece. Once the pocket is sewn, it

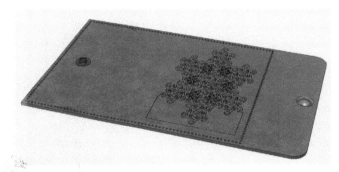

FIGURE 5-37: Extended cover

will cover up the interior stitches of the tab. This is great for making things look good, but it's frustrating when you forget and discover you have to cut the stitches and remove a piece because you forgot to sew things in the right order.

ENHANCEMENTS **87**

THE GHOULS AND GEARS MULTI-TOOL HOLDER

Modern multi-tools typically come with a nylon belt holder. These ballistic nylon cases always seem to wear out faster than I think they should. A leather belt holder is an easy replacement and offers lots of opportunities for customization.

DESIGN

THE BASIC LEATHER PATTERN

You can design the pattern for the belt holder in a variety of ways. The approach we'll take in this chapter is to use a single piece of leather that is riveted together. It's appealing to use two pieces of leather, one to surround the tool and the other to form the belt holder and top flap, but I've chosen to use a single piece to help illustrate some pattern design principles.

The biggest downside to a single-piece approach is that, unless you're making multiples, it doesn't make the most efficient use of leather. The cut pattern is roughly T-shaped, which nests well if you're cutting multiple pieces, but leaves some unusable areas when you're cutting a single piece. I've found that I usually end up finding uses for the leftovers, so I tend not to worry much about this.

Multi-tools vary in size and shape. Most fold up into a rectangular cube of sorts, but the dimensions of that cube can be very different depending on brand and model. I have owned dozens of different multi-tools, chasing the perfect combination of fit and function. Currently, I have settled on the Leatherman Wave® as the tool I carry for everyday. (See Figure 6-1.)

This tool, with some exceptions, is 30 mm × 100 mm × 17 mm when folded. The exceptions are the ridges of the knife blades, which rise up to 4 mm above the sides. This gives the tool a 38 mm width, though it is uneven. (See Figure 6-2.)

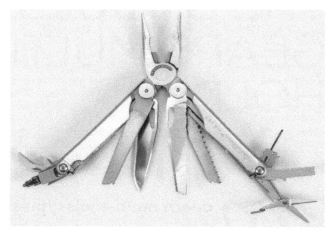

FIGURE 6-1: Leatherman Wave®

FIGURE 6-2: Folded view of the tool

You can take several approaches to securing the tool with the holder. I've made friction-fit holders for other tools that were very successful, but for this project we'll use a top flap. The irregular profile of this tool does not lend itself to a friction-fit design. Since the flap will keep the tool from falling out, we'll make the interior just slightly larger than the tool, with dimensions of 40 mm × 100 mm × 20 mm. This will make it easy to retrieve the tool without tugging.

The other major design decision is how to fasten the leather to itself. We hand-stitched in the last chapter and that is an excellent choice. In this chapter, I want to introduce one of the other major methods of fastening used in leatherwork: riveting. We will use four rivets to secure the belt loop section to the two sides in the back. A line 20 snap will allow the top flap to close the case securely.

There are a few grace notes that are worth including in the design. It can be very frustrating when you want to quickly retrieve your tool, but can't get an easy grip on it. To address this, I've added a semicircular cutout on the front panel. Two quarter circles on the back panels form a matching semicircle once the rivets are in place. This provides an easy way to pinch the top of the tool to get a grip on it.

Setting rivets deep inside a narrow case can also be a challenge. I've made cutouts on the bottom of each side to allow access to the lower rivets. Additional fillets on various edges help reduce sharp corners, which tend to wear poorly.

The last shaping issue relates to our motif. The top flap is designed to end in a skull head. Since the top flap also includes a snap, you must carefully position the snap in the middle of the cranium, rather than blocking the eyes or nose. I chose a skull that didn't have a jaw, because I didn't want to have too much material below the snap on the flap. Over time, with wear, this would have eventually started to curl, since it would be the place that is most frequently grabbed and pulled when opening the case. The cut pattern for the end of the top flap is now the outline of the lower part of the skull. (See Figure 6-3.)

THE DIGITAL IMAGERY

The other component of our design motif is gears. I'm not sure why I find that gears and skulls go together, but it's a motif I've come

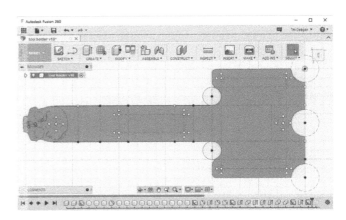

FIGURE 6-3: Basic cut pattern

back to many times over the years. Maybe it's a steampunk thing, or maybe it's that both represent the internal hidden structure of things. Either way, it works for me.

In the DXF, we used the skull image to create the outline of the skull perimeter at the end of the flap, but we'll need to overlay the skull in the burn pass as well to get detail.

I want gears to show everywhere except where the skull is positioned. I also want the burn time on the laser to stay in a reasonable range, so I want to limit the position of the gears as much as possible. I'll avoid putting gears on the area between the rivets in the back and on the back of the two sections that join inside. (See Figure 6-4.)

The skull isn't as detailed as some of the gears, but we'll make up for that with some leatherworking dye techniques later in the chapter.

FIGURE 6-4: Gear and skull burn pattern

Leather's Irregularities

Leather teases and delights with its inconsistencies. It's reasonable to expect that pieces cut from a section of leather that are close to one another will behave similarly, but even pieces cut slightly farther away can vary in thickness or finish. Many sections of leather have brand marks or scars on them. While imperfections like these can be frustrating, they are also worth taking into account for their aesthetic or design possibilities.

I mention all these variances and imperfections to remind you that it is critical to test and retest the settings chosen for cutting leather with digital fabrication tools. Variations in thickness can result in incomplete cuts. Use scrap pieces from close to the area you're intending to cut from to test your intended cut and burn settings.

DIGITAL LASER FABRICATION

PREPARING THE FILES

Once again, we need two files to fabricate this project. Tool_Holder_CUT will be a DXF file and Tool_Holder_BURN will be a PNG image file.

The intent is to create a piece of leather that is wide enough to wrap all the way around the multi-tool, and has a long enough flap to wrap under the tool, up the back as the belt holder, and over the top to snap onto the front. Since the space we're shooting for is 40 mm × 100 mm × 20 mm, the width would be double the width of 40 mm and double the thickness of 20 mm. Added up, this comes to 120 mm to allow enough leather to wrap around the tool. For the other direction, we start with the 100 mm for the front, 20 mm for the thickness, another 100 mm up the back, 20 mm over the top, and then 40 mm for the flap. (See Figure 6-5.)

Assembling the shape for the pattern was accomplished in Adobe Fusion 360 using geometric primitives such as cubes and cylinders. I used these primitives to form more elaborate shapes by adding and subtracting them from one another. Many tools refer to this as *constructive solid geometry (CSG)*. Other tools will refer to the addition or subtraction of shapes as *Boolean operations*. (See Figure 6-6.)

In addition to the outline I wanted to cut, I subtracted the holes for the rivets and snaps from the model. I used several *construction*

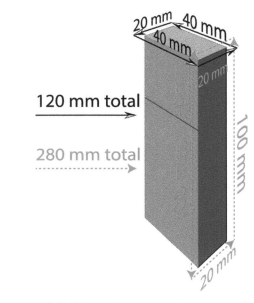

FIGURE 6-5: Dimensions around and over the tool

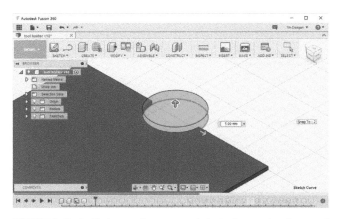

FIGURE 6-6: Subtracting a semicircle from the front of the holder

lines in Fusion 360 to help me position these correctly. Construction lines do not interact with the model, so they don't modify the shapes they are positioned on. (See Figure 6-7.)

The skull on the end of the flap is more complex than I like for a CSG approach. In this case, I searched the internet for a skull image that I liked and traced it to create a vector image. (See Figure 6-8.)

I deleted the points on the lines that made the lower jaw to remove it from the image. I could have done this by erasing a section of the raster image prior to tracing; it's really a matter of preference as to which approach you take. (See Figure 6-9.)

Once I had the skull the way I wanted it, I saved it as a DXF file, which I imported into Fusion 360. I needed to move it to the end of the flap and combine it with the rest of the object. It was important for me that the snap was positioned in the middle of the skull (rather than on top of the eyes or supraorbital ridge below the eyebrows). I made a circle with the diameter of a snap centered on the hole I created for the snap and made it a construction line. This allowed me to get a clear idea of where the snap was in relation to the skull. (See Figure 6-10.)

With the 2D sketch positioned correctly, I used a process called *extruding*, which takes the 2D shape and pushes it into a third

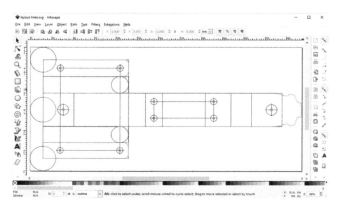

FIGURE 6-7: Positioning the holes with construction lines

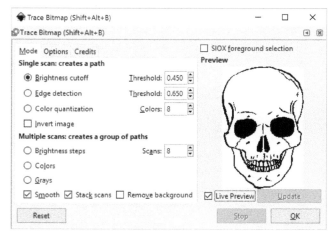

FIGURE 6-8: Tracing the skull in Inkscape

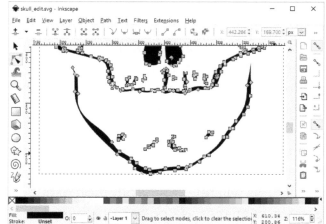

FIGURE 6-9: Removing the jaw

> ## *Cleaning Up Traced Vector Files*
>
> Tracing functions are frequently merciless. They have no concern for how fussy and complex the resulting vector image is. What might have worked fine with 50 points ends up having 500. Many programs provide parameters to modify the tracing function to various degrees, but inevitably you end up with too many points. This is especially challenging when it results in tiny overlapping loops. Extruding these can cause many bizarre or frustrating outcomes.
>
> Most programs such as Inkscape or Adobe Illustrator provide functions to "simplify" a vector image. This can help to a degree, but can also degrade the image beyond usefulness if taken too far. Ultimately, editing the vector by hand to clean up redundant or confusing details is essential for getting useful results when using these files for digital fabrication.

dimension. In this case, I extruded it 2 mm so that it became a 3D object the same thickness as the CSG object I'd made for the rest of the pattern. (See Figure 6-11.)

Once the 3D version of my pattern was complete, I selected one of the faces to create a 2D sketch. I saved this sketch to Tool_Holder_CUT.DXF. (See Figure 6-12.)

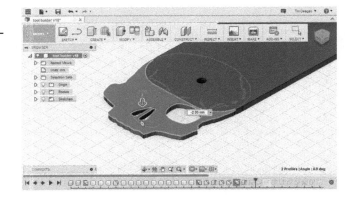

FIGURE 6-11: Extruding the skull

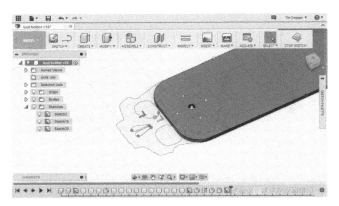

FIGURE 6-10: Positioning the skull in relation to the snap

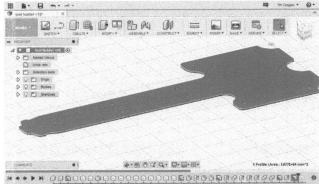

FIGURE 6-12: 3D object with a 2D face

DIGITAL LASER FABRICATION

The burn pattern doesn't have to fit the cut pattern exactly. Burning some of the area that will be cut away is no big deal except for one thing: time. Frustratingly, an image this size, operating at 1200 mm/min in raster mode with the 2.5 W laser I'm using can take almost 4 hours to burn at 10 lines per millimeter. To cut this time down, I'll do two things. The first is to make sure I'm not burning any place other than the area I want, and the second is to convert the image from raster to vector using a trace function. (See Figure 6-13.)

I'll use the free graphic editing tool, Inkscape, to position and mask the images.

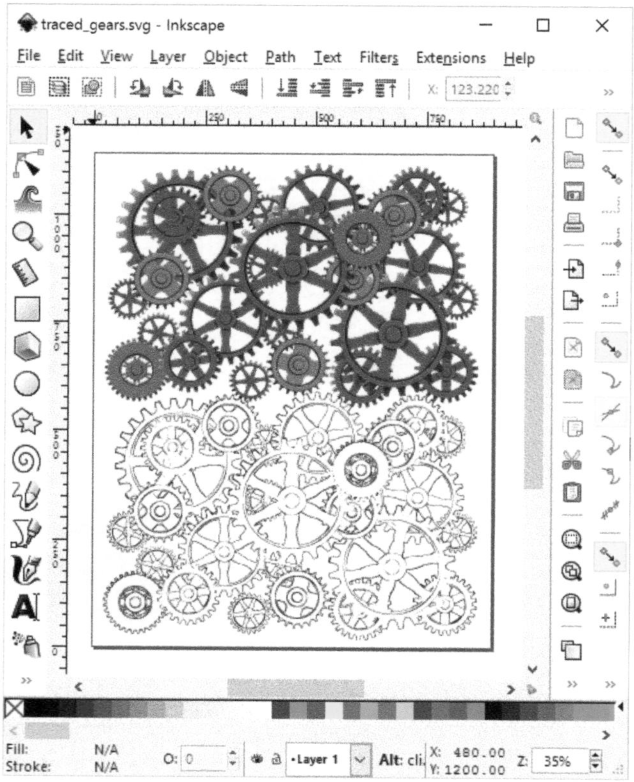

Inkscape can load the DXF files created in Fusion 360 and use them to clip the areas we want covered. Inkscape is also an excellent tracing tool.

To start with, open the Tool_Holder_CUT.DXF file created earlier and make it its own layer. Unfortunately, the DXF import is likely to result in a set of unconnected line segments, rather than a single object. If, when you select the import and give it a fill color, you see something like the image below, it will be necessary to manually join the line segments. (See Figure 6-14.)

Select the end nodes of two adjoining lines and click the Join control to connect them. (See Figure 6-15.)

Once this is complete, you could go on to subtract the rivet and snap holes from the object so they aren't part of the clipping mask. The holes in this project are small enough that it doesn't really make a lot of difference if they're there or not, but on other projects, you might find it important.

FIGURE 6-13: Vector trace of the raster burn pattern

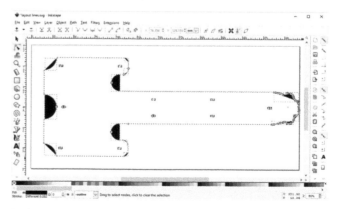

FIGURE 6-14: Unjoined line segments

FIGURE 6-15: Joining two line segments

Check your work by adding a fill to the object. The fill should be inside the bounds. (See Figure 6-16.)

On another layer below the outline, place the gears and scale them so they fill the desired areas. Trace them to turn them from bitmapped objects into vector objects. In Inkscape you can only clip vector objects, and we will use the pattern outline as a frame to clip with. Clipping allows you to use an image as a window to show or block out parts of another image.

We could have done a similar set of operations in Photoshop or GIMP and masked the area, but neither of those tools happily imports DXF files. Once you have vector versions, copy the outline object from the layer it's on and paste it into the layer with the gears, arranging it to be the top object. (See Figure 6-17.)

Select the gears and the outline and then right-click and choose Set Clip, or select Object ➪ Clip ➪ Set. This should now only show the gears through the area bounded by the outline. You can hide the layer that still has a copy of the outline. The image should look like what you see in Figure 6-18.

The only remaining task is to import the skull into a new layer and scale and position it so that it appears directly above its

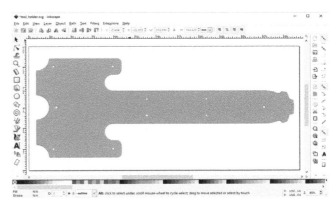

FIGURE 6-16: Filled pattern outline

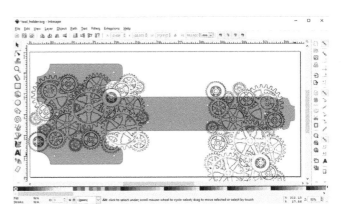

FIGURE 6-17: Gears positioned under the outline

DIGITAL LASER FABRICATION

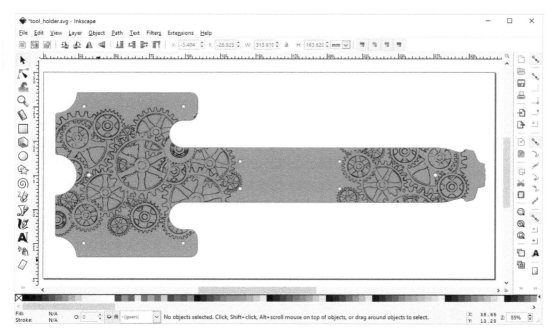

FIGURE 6-18: Clipped gears

location in the outline. If the skull allows the gears to show through, you'll need to make a second copy of the skull and click Path ➪ Break Apart. This will create a filled-in black skull. Without unselecting the black skull set of objects you just made, hit Ctrl+G to group the objects together, and then change the fill color to white. We'll use this white skull below our first transparent skull image to hide the underlying gears. To do this, use the Arrange menu to place the white skull group below the transparent skull and then orient it so that it's positioned perfectly beneath the first skull. Ultimately, it should all look like Figure 6-19.

Make sure you've set Document Properties to Resize Page to Drawing or Selection so that the final dimensions of the page are the same as the DXF. The skull makes it a little longer than our original 280 mm; in this case it comes to 302.9 mm. The width should remain at 120 mm, as we planned.

Export the image as a PNG: Tool_Holder_BURN.PNG. Perhaps you are thinking, "Hey! Why go back to raster?" This is an excellent question. The reason is that DXF exports tend to be . . . let me think of a polite

FIGURE 6-19: Finished burn pattern

term . . . weak. Among other problems, exporting the DXF from Inkscape tosses out the clipping and shows the entire set of gears. It's much easier to export a bitmap, import that into the control program, and trace again. Figure 6-20 shows the traced result from T2Laser.

With all that complete, we have our cut and burn files ready to go.

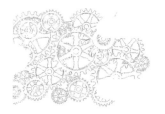 

FIGURE 6-20: Re-traced gears and skull

Handling Smaller Work Areas

This project has a length of just over 300 mm. Many desktop laser engraving tools have a work space with dimensions that are smaller than this. Dimensions of 130 mm × 200 mm and 300 mm × 200 mm are common on other machines. The smallest machines commonly sold have a 38 mm × 38 mm area, but these are too small to consider for practical leatherwork. (See Figure 6-21.)

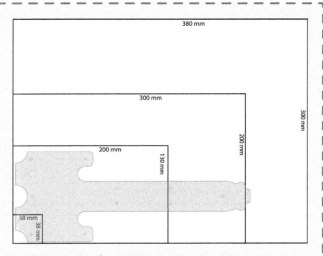

FIGURE 6-21: Common laser tool work areas

The other two sizes are very usable, but may require some modifications to the files. For the 300 mm × 200 mm beds, common on 40 W lasers, rotating the images to fit is a viable approach. (See Figure 6-22.)

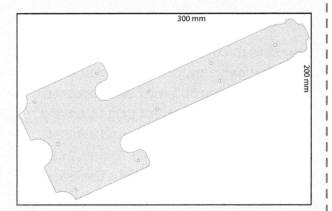

FIGURE 6-22: Rotating the file

DIGITAL LASER FABRICATION

For smaller beds, such as the common 200 mm × 130 mm space, cutting the piece into two smaller pieces is required. Always look for an existing join spot, such as the two lower rivets on the back, where the two pieces could overlap and be connected. (See Figure 6-23.)

It's also important to do a rail-to-rail test of your machine. Quoted working areas are frequently over-optimistic. The 380 mm quoted on my machine ends up actually being 360 mm in real life.

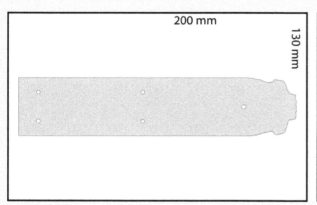 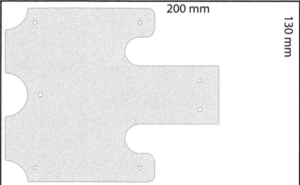

FIGURE 6-23: Overlapping sections

PREPPING THE WORKPIECE

We need to make the cut from a piece of leather that can support the needed dimensions of 120 mm × 303 mm. As discussed in the previous chapter, it's best to have a little overlap, so the piece should exceed these dimensions by at least 10 mm in each direction. Before you rough cut the leather for this, consider if you might want to make more than one holder (now or in the future).

Since the cut pattern is roughly T shaped, it makes better use of the leather if you nest at least two holders in the rough cut. (See Figure 6-24.)

FIGURE 6-24: Nested patterns

Orient the rough cut leather in your tool so that the entire pattern is within the reach of the laser. I like to cut wet leather, but you can choose to cut dry or wet. Remember to adjust your feed rates (and possibly power) for the method you are choosing to use.

If you're working with wet leather, give the leather an even soaking. This is a lengthy pattern to cut and burn, so you may need to mist the leather occasionally if it starts to dry and lift off the work surface. As always, if you're placing wet leather on top of an MDF or other surface that will warp when wet, lay down a piece of wax paper under the leather to avoid getting your waste board wet.

READY TO CUT

Import Tool_Holder_CUT.DXF into your program and verify the dimensions. Set the power and feed rate to the values you decided on. As noted earlier, you will want to do a test cut on the leather to verify that parameters are still working as desired. I ended up using the full power of my 2.5 W laser and doing two passes at 40 mm/min for the cut. If you don't move the workpiece, it's perfectly appropriate to do multiple cut passes. (See Figure 6-25.)

Once you finish your cut, do not move the leather. The engrave (burn) pass relies on the registration of the leather not having changed.

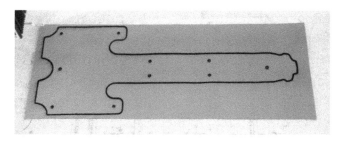

FIGURE 6-25: The cut leather pattern

READY TO ENGRAVE

Import Tool_Holder_BURN.PNG and verify the dimensions. Many programs confuse units and will determine the size of the image by taking the pixel dimensions and dividing it by the resolution value. If your image is 1145 pixels wide and the resolution is set to 0.1 (10 lines per millimeter), then the program may assume the image is 114.5 mm wide. If you really want the width to be 120 mm, then change the dimensions to reflect this.

For speed's sake, I prefer to burn in vector mode. As noted above, I use the auto-trace function to convert the image. My feed rate was 1200 mm/min, but I probably could have pushed this a little faster. On a more powerful laser, higher feed rates are more viable.

Send the file to the laser controller and burn the pattern. (See Figure 6-26.)

CLEAN UP AND TRIM

If you didn't completely cut through the leather at all points on the outline, it's more

DIGITAL LASER FABRICATION

effective to let the leather dry out before trying to remove it. The dry leather will often allow you to snap the piece out of the surrounding leather. You can also use a utility knife to cut the last remaining bits of leather holding the piece in place.

To minimize warping, it helps to hang the leather on a line to dry. Remember to rinse it first to remove as much of the carbonized material as possible.

It's also easier to clear out the holes when the leather is dry, and a small pair of embroidery scissors is great for clearing the hole and cutting off any dangling pieces of the interior.

The utility knife or embroidery scissors can be used to trim off any rough leather

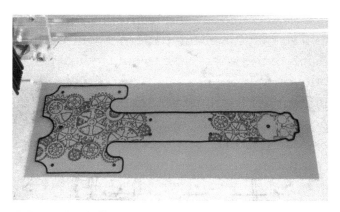

FIGURE 6-26: Burning gears

fibers that surround the rim. It's desirable to have the edges be as clean as possible so that you can easily dye any of the lighter parts where you had to cut.

LEATHERWORKING

We'll use two leatherworking techniques on this tool holder. The first will be using a resist-dye process to highlight the skull in relation to the rest of the holder. The second will be riveting the piece together.

COLOR AND FINISH

Since the laser did us the favor of darkening (i.e., burning) the leather where we engraved the gears and skull, we don't have to do anything to highlight them. But most people prefer a darker tone to leather than the undyed leather provides. We'll apply dye to achieve this, but not everywhere.

The skull is a distinctive element that we want to emphasize. Bones are also typically lighter than a warm brown leather tone like the Canyon Tan dye I intend to use on the rest of this piece. Additionally, I'd like the eyes and nose of the skull to be very dark to suggest the skull is hollow. We could achieve this effect several ways.

It's perfectly reasonable to simply hand-paint the dye onto the leather and just use different colors (or no color at all) to achieve the different effects we're looking for. But I'm hooked on using an airbrush to get an even covering of dye. Brushing or sponging dye on dry leather can be frustrating. Unless you're extremely careful, the dry leather can soak up too much of the dye when the brush touches it, creating an overloaded spot. These soaked spots can leave a mottled finish. The airbrush avoids this by providing an even spray. (See Figure 6-27.)

The downside to airbrushing is that it is difficult to avoid over-spraying and getting dye in undesired areas. The solution is to cover areas that you don't want dyed with a *resist*. This is a finish product that seals the leather and allows you to wipe off any dye that lands on it without coloring the leather underneath.

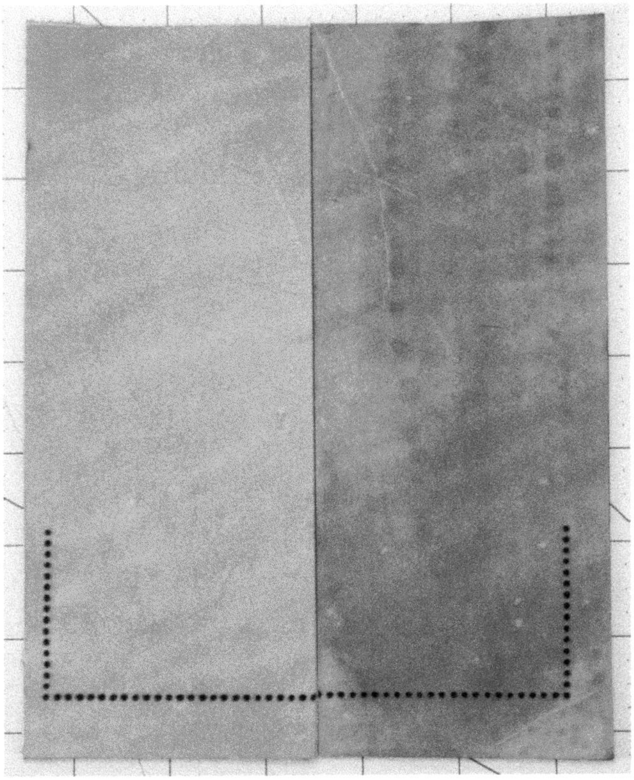

For this design, I will paint a resist onto the areas of the skull that I want to keep light. I want to leave them the color of the undyed leather, so I'll do it before starting any airbrushing. It also works to spray everything with a light coat or two of dye before adding resist if you want the leather slightly darker.

I'm using Eco-Flo Super Shene, a Tandy product, which can serve as a resist or finish. A small #3/0 brush allows me to apply the resist selectively to the desired areas. If you accidentally apply resist to an area, you've got a very brief moment to wipe it off before it soaks into the leather. Keep a wet paper towel on hand for accidents. (See Figure 6-28.)

Brush the resist evenly and allow it to dry. For best results, apply a second coat and allow it to dry. Once it's fully set, you can airbrush or sponge the dye onto the rest of the holder. It's worth noting that, just as

FIGURE 6-27: Airbrushed vs. badly brushed dye

FIGURE 6-28: Applying resist to the skull

successive coats of dye can be applied to get darker tones, you can apply resist to other parts of the model so that they stop darkening with each coat. I chose to add resist to the areas between the gears after a coat of dye so that the gears darkened the most. (See Figure 6-29.)

The important consideration when using a resist is that once you have applied it to an area, you can no longer color it. The sequence of operations is very important; you should plan your steps carefully. We'll use another resist technique involving digitally cut adhesive vinyl stencils in a later chapter that gets around this problem.

Once your dye is applied and dry, buff it with a clean cloth or some clean sheep's wool. Afterward, apply an even coat of resist (either more Super Shene or another product) to the entire piece.

SETTING THE SNAP

There are many ways to keep a flap closed. Magnets, studs, and buttons all have their places, but the most common approach is a snap. Snaps come in different sizes, colors, and designs. The most common types are line 20 and line 24. Line 24 snaps have a post diameter of 5/16" with a snap diameter of 9/16". Line 20 posts are 3/16" and have a 7/16" snap. There are also less common, smaller line 16 snaps that are often referred to as *glove snaps*.

Snaps have four parts. A *cap* that attaches to a *socket*, to form the female top, and an *eyelet* that attaches to a *stud*, to form the male bottom. The cap and the eyelet both have posts that pass through a hole in the leather and are set with a tool to hold the socket or stud in place. (See Figure 6-30.)

FIGURE 6-29: Applying resist to additional areas

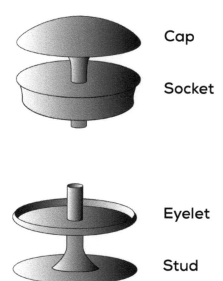

FIGURE 6-30: Anatomy of a snap

The lighter line 20 snaps work well for our project. These will work with the holes we laser cut into the leather on the flap and body. We'll set these before we rivet the body so that we have easy access.

Starting with the body, push the eyelet through from the back. Place the stud over the eyelet and place the base of the eyelet on a hard surface that can withstand hammering (don't do it on your nice kitchen counters). Traditionally, this surface would be an anvil.

Actual anvils are ideal, but have become startlingly expensive and increasingly difficult to acquire. Heaviness provides stability, which is valuable, but the critical aspect that is needed for this task is a hard surface that doesn't allow the item being hammered to move or get off-kilter. Other solutions could be the flat surface on the back of a vise, or a section of metal stock. Just make sure that the surface is flat and you can keep it stable while hammering on it. (See Figure 6-31.)

Position the setting tool directly on top of the post and be sure to hold it in line. Do not allow the setting tool to tilt while hammering or the post will be set off center. On the stud, this might not be a deal-killer, but on the socket an off-center post can get in the way of allowing the socket to fit on the stud.

Leather stamping and setting tools are intended to be hammered with a non-metal hammer head. Many leatherworkers prefer to use a rawhide mallet. I currently use a hammer wth a hard plastic head. Whichever you choose, hammer the other end of the setting tool until it is flush with the leather. The tool should be held vertically so that it does not veer off to one side or the other and set the rivet asymmetrically.

Hammer the post down so that it is as flush as possible with the inside of the stud. (See Figure 6-32.)

FIGURE 6-31: Traditional and impromptu anvils

FIGURE 6-32: Setting the stud

The operation on the top flap is fundamentally the same, with one important distinction. The cap is convex and would be hammered flat if it were set directly on the anvil. However, the setting tool should have come with a disk with one flat and one concave side. This is also referred to as an anvil and is placed on the primary surface with the cap resting in it. This allows you to hammer on the cap's post without flattening the cap. Push the cap's post through the leather, place the socket on it, and use the setting tool to hammer the post down to hold the socket in place. (See Figure 6-33.)

That's all there is to setting a snap. It can be frustrating the first few times you try it, so I highly recommend doing a few practice runs on scrap before working on a piece you care about.

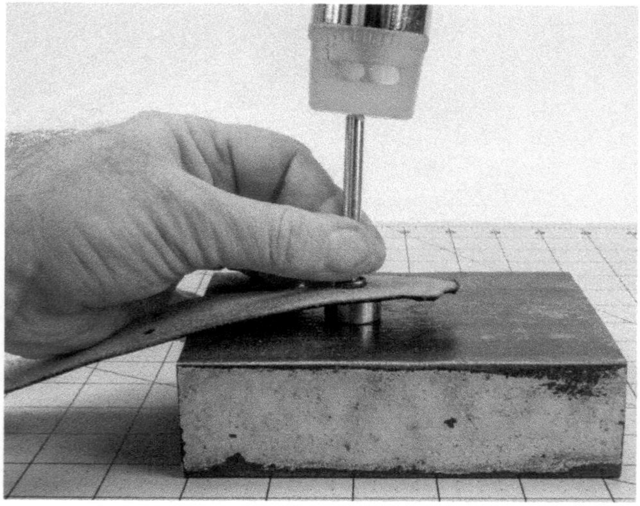

RIVETING

I prefer to do my dye and resist work prior to construction (riveting or sewing). This allows me to get an even coating, including under the attached pieces. Like many things in leatherworking, this isn't the only way to do things. Some people may prefer to rivet the piece first and then apply the dye and/or finish, and there may be instances in which it's better.

We'll use tubular rivets for this project. Tubular rivets are strong and easy to set. Their biggest downside is that they produce a set of star-shaped tines on the back that can occasionally have light burrs (see Figure 6-34). You wouldn't want to wear the back side against skin, or have it against a softer material. It may be tempting to

FIGURE 6-33: Setting the socket

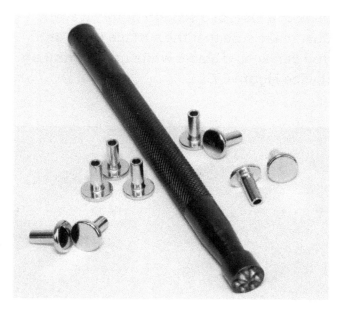

FIGURE 6-34: Tubular rivets

THE GHOULS AND GEARS MULTI-TOOL HOLDER

hammer these flat, but this usually overstresses the tines and they break off with wear, causing the rivet to fail.

Our biggest challenge is the pair of rivets at the bottom of the holder. We need a strong, flat surface the rivet face can rest on while we hammer the setting tool against the other side. Since the rivets are deep inside the case, it's tough to get anything far enough inside to serve as an anvil. I usually use a piece of box steel clamped to my grandfather's 150 lb. anvil. But this isn't an easy setup to duplicate, so I've made modifications to this design to make the problem easier to solve.

We'll rivet the back sides of the case to the belt loop that wraps up the back and over the top to form the flap. There are four rivets to set in the back; we'll use a snap on the front flap, so don't rivet that.

You probably noticed the holes at the bottom of the holder's sides. These were placed to allow a small section of ½" bar stock to pass through underneath the rivets. The ends of this stock can be set on a couple of pieces of 2×4 or anything else that will solidly support both sides.

Push the tubular rivet in from the inside so that the flat face is inside the holder. Place the setting tool so the dimple in the middle fits into the end of the rivet tube. (See Figure 6-35.)

Hammer the setting tool straight down so that a set of tines form evenly and bite into the leather. In the worst case, you can use needle-nose pliers to break the tines of the rivet off, pull it out, and try again. But these rivets are relatively easy to set, and you should end up with a solid hold. (See Figure 6-36.)

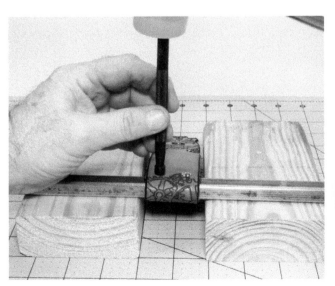

FIGURE 6-35: Setting the bottom rivets

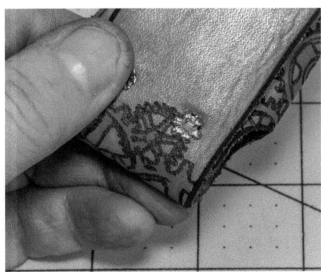

FIGURE 6-36: Correctly set tubular rivet

With the bottom two set, go ahead and set the top two rivets. When placed diagonally, the bar used to set the bottom rivets will work for the top ones, as well. (See Figure 6-37.)

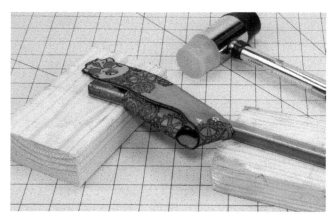

FIGURE 6-37: Setting the top rivets

THE RESULTS

The resulting multi-tool holder is a functional, unique eye-catcher! (See Figure 6-38.)

FIGURE 6-38: The final results

STEAMPUNK ACTION-CAM TOP HAT

S teampunk has a vast number of aficionados around the world. The alternate-future past aesthetic, where the raciest Victorians and Edwardians ride airships and drive giant steam-driven walking machines, captures many people's imaginations. Leatherworking is a mainstay of the Steampunk fashion toolset, and the ability to leverage digital fabrication techniques feels perfectly appropriate for the "tech-before-its-time" style.

DESIGN

Gears are a consistent motif in many Steampunk outfits. These may be harvested from watches, old mechanical equipment, disassembled bicycles, and other sources, or (surprise!) 3D printed. Printing your own gears and other adornments allows custom sizing and mounting potential that dramatically enhances their usability in costume design.

My choice for a Steampunk item is an action-cam holder. It didn't take me long to decide where an intrepid gentleman of science would mount his action cam. Obviously, it would be on his top hat (for reasons obviously having to do with me wanting an excuse to buy a top hat).

I own an ELE Explorer action cam that is the same size and shape as the GoPro Hero. The same processes I'm using can be adapted for any action cam. We will start with taking careful measurements of the camera and building a 3D model. This model will serve us well for both the design of the holder and to produce a physical model we can use for fitting wet leather. Like many items that you would build leather holders for, it is undesirable to get uncased action cams wet.

TAKING MEASUREMENTS

Few tools are used as often during digital fabrication as the humble caliper. The modern digital readout caliper is a delightfully easy way to measure things. A basic 6" digital caliper from even the cheapo import tool store is indispensable. Calipers can cost up to thousands of dollars, and for some purposes that is money well spent. But for hobby-level 3D printing, CNC, and leatherwork, basic models with metric and inch measurements costing less than $30 are perfectly useful. I own multiple pairs so that they're always at hand. I am from the "affordability = access" school of tools. If you're from the "buy only quality" school, know that you have my respect—and can you leave me your tools in your will?

Note the overall dimensions of the unit, and then begin measuring the feature sets. For our purposes, it's not critical to be accurate to more than a tenth of a millimeter. We'll want to determine the center point of each circular feature, so measure the diameter, and then subtract the radius from the distance to the edge. (See Figure 7-1.)

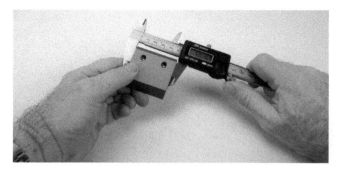

FIGURE 7-1: Measuring with digital calipers

I've made a number of holders for my cam. I originally created the model for it in the wonderful OpenSCAD. It's very easy to be precise in OpenSCAD. In this model, I adjusted the height of the features so that I could see them protrude through other models. That way, I could be sure that holes made to get at the buttons or lens would align. (See Figure 7-2.)

Luckily, it was easy to use the resulting STL in other CAD programs. I imported the STL into Fusion 360 and used the Mesh to BRep functionality. This feature can be hard to discover on your own; check Google for good videos on the topic. Ultimately, the Mesh to BRep function will let me convert the triangles of the STL mesh into a body object that can interact with other body objects in Fusion 360.

I wanted to cover the front, top, and sides of the camera with decorated leather joined to a hatband that wrapped around the hat. Measurements of the target hat gave me the length of the hatband arms; they didn't need to meet, since I wanted to use some grommets and a little corsetry in the back to tie the project in place. (See Figure 7-3.)

The hatband provided a nice surface for some laser patterning, but I wanted something more for the front. Clearly, I need gears.

There is no way that I'm putting a gear on something unless it at least moves. An immobile gear is a crime I cannot willfully commit. I decided that I could put one large gear onto the lens and a smaller one on the front button. I'd need to mount these so they spun.

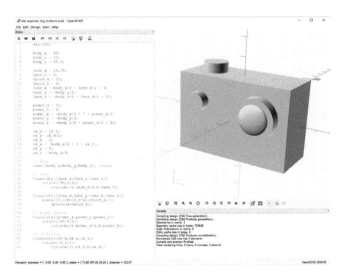

FIGURE 7-2: OpenSCAD model of the action cam

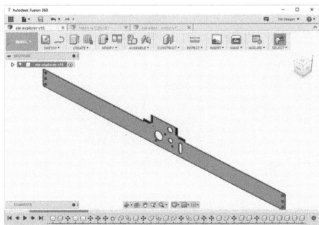

FIGURE 7-3: Hatband model

THE BASIC LEATHER PATTERN

I used the model I'd made of the camera to test the positioning of the holes on the leather pattern. This was indispensable for getting the lengths and positions correct. I used multiple cameras in different rotations to check all sides before folding. (See Figure 7-4.)

It became apparent that I wouldn't be able to reach the front button with the leather, mount, and gear in front of it, so I also added a button extender. (See Figure 7-5.)

I used an old sewing pattern–maker's trick to test this model. Using 2 mm craft foam, I did a quick cut of the main part of the holder and made the desired folds to test if the holes matched up. (See Figure 7-6.)

For the burn pattern, I wanted something with an industrial motif, and found a border pattern in an old book of copyright-free designs (*Decorative Frames and Borders, 396 Copyright-Free Designs for Artists and Craftsmen*, selected by Edmund V. Gillon Jr., 1973, Dover Publications). These types of books have fallen out of favor with the advent of the internet, but they are cheap at used book stores and are absolute gold

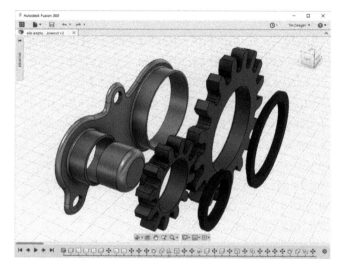

FIGURE 7-5: Gear model

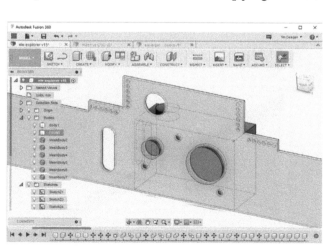

FIGURE 7-4: Camera model used to check leather model

FIGURE 7-6: Pattern test

when you're trying to find designs to burn, etch, engrave, or carve.

Scan images into the computer and use the trace function in Inkscape or a similar program to turn them into DXF files that can be used in digital fabrication. (See Figure 7-7.)

I added the traced image to the hatband sides for the final leather pattern. (See Figure 7-8.)

FIGURE 7-7: Scanned border pattern

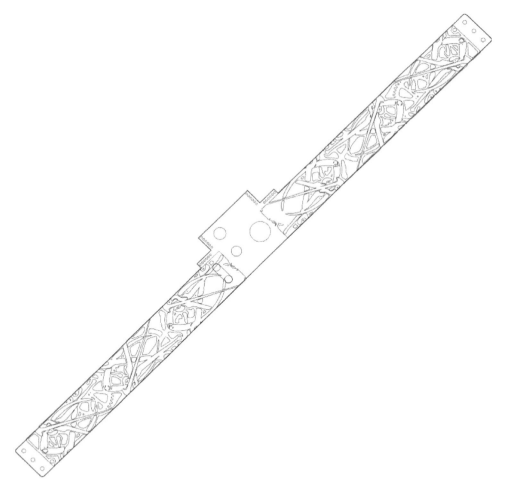

FIGURE 7-8: Complete leather pattern

DIGITAL FABRICATION

This project uses a 3D printer and a CNC-controlled laser. Because the length of the workpiece exceeds the size of my smaller laser's work space, I'm using a 15 W laser attached to my 1000 mm × 1000 mm CNC machine, rather than the gantry laser I used in an earlier project. This also helps demonstrate that the activities occurring at different power levels and sizes are basically identical.

Any 3D printer can perform the printing tasks. I'm using a Prusa i3 MK2, but there is nothing in this project that any machine would have trouble with. I'm printing with PLA, though any filament type will work, as long as the thin walls of the gear-holder will be strong and not break or delaminate. This was a problem with my experiments using bronze-fill filament—thin cylinders snapped right off. Make a test print with the filament of your choice and abuse it to make sure it will hold up under wear.

PREPARING THE FILES

There are three DXF files for the laser work and two STL files for the 3D printer. The DXF files are tophat_outline.DXF, tophat_holes.DXF, and gear_plate.STL, meaning that it combines multiple models into a single file. The other is camera.STL.

The laser files are created just as in the multi-tool holder project. You'll want to design your own to fit the size of your top hat and camera. I separated the outline and hole-cutting files so that I could be kind to the duty cycle of my laser. Separating the cutting files also allows choices. The outline cuts are simple enough that they could be done by hand if desired. The burn file is able to be burned lightly and quickly, since it's really just a method to transfer a pattern to the leather for painting.

I had to mount the workpiece diagonally to get it to fit comfortably on my work table. I measured the width required and determined that I needed a piece of leather 3" × 27". Having drawn a diagonal line on my spoil board, I marked parallel lines 1½" inches to either side. This allowed me to position the blank where I needed it. The DXF files were prepared so that the cuts and burn would be made in the appropriate locations. As an extra check, I lightly burned the pattern onto tape I placed diagonally on the spoil board to make sure everything would fit. (See Figure 7-9.)

PREPPING THE WORKPIECE

Hatbands don't need to be very heavy, so I used relatively light 4–5 oz leather. I cut the necessary 3" × 27" blank from a single shoulder of veg tan using a rotary knife. A utility knife, head knife, or any other sufficiently sharp tool would have worked, including good leather shears. (See Figure 7-10.)

Since I prefer to cut wet leather, I lightly soaked my blank, and then placed it on wax paper atop the spoil board using the marked lines to register the piece. The wet leather laid flat and, as long as only photons from the laser were touching it, didn't need any special workholding (methods to hold it down). If I'd used a drag knife or router bit, I would have needed to secure the leather in place. (See Figure 7-11.)

LASER CUTTING AND BURNING

To get started, I'll use the following laser files:

1. Tophat_cut_outline.DXF
2. Tophat_cut_holes.DXF
3. Tophat_burn.DXF

I created the G-code for the tophat_cut_outline.DXF file with T2Laser, using 70 mm/sec for the feed rate and 255 (full) for the power. I ran the pattern twice, pausing at each end of the workpiece to allow the laser to cool. You'll need some time and experience with your setup to determine how long to run without breaks. It's worth noting that you can use a utility knife to finish the cutting if you can't burn all the way through. The initial burns make an excellent guide.

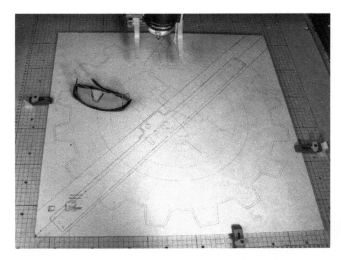

FIGURE 7-9: Pattern check on the spoil board

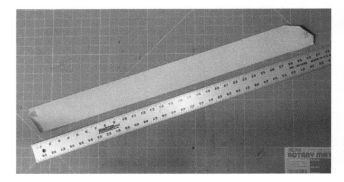

FIGURE 7-10: Work blank

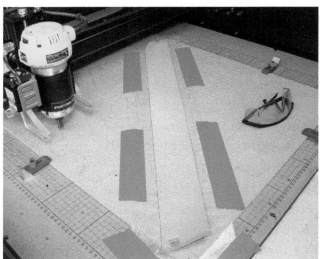

FIGURE 7-11: Leather placed for cutting

The Tophat_cut_holes.DXF file was also loaded into T2Laser and converted to G-code with the same settings as the outline file. It's helpful to get full burn-through on the larger circles, but pay attention with the smaller ones. If the charring gets excessive, you'll lose the necessary leather between the holes to allow stitching.

The Tophat_burn.DXF file is loaded into T2Laser and is given a much higher feed rate. I went up to 1500 mm/min, since I only wanted the pattern transferred for later blocking and painting.

With all three files complete, the leather is ready for manual leatherworking. (See Figure 7-12.)

MAKING THE GEARS

As noted earlier, I'm printing with PLA, but there is nothing special about it. My experiments with thin-wall bronze-fill were a fail, so while it would be great to print metallic-looking gears, I'm settling for plastic with metallic enamel model paint.

The gears that will be displayed will have their bottom layer facing the world, so it really helps to make sure you have good adhesion and a good-looking first layer. If this is a problem, you could always mod the gears so they print in two parts: standoff ring and gear. These are combined on the current model. The standoff ring serves to keep the gear above the heads of the rivets. Otherwise, they won't spin well.

I'm printing at 0.15 mm layers to get good-looking prints. Many parts are fairly thin, so tight layers help the accuracy, as well. It took me about three hours and 40 minutes to print the full plate. (See Figure 7-13.)

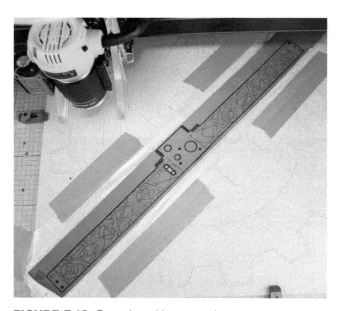

FIGURE 7-12: Completed laser work

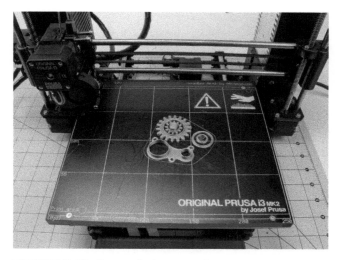

FIGURE 7-13: Gear printing complete

Once the print is complete, give it time to cool (if you used a heated bed) so the print doesn't get warped during removal. You may need to give the parts a light sanding with fine-grit sandpaper to get the gears to spin easily on their shafts. I've found it a little easier if you use slicer settings that randomize start positions. It's harder to sand off a ridge on some of the thinner parts.

The gears will be held on the shafts with small retaining rings. Once the base has been riveted to the leather (in a later step), the gears can be placed on the shafts, and a very tiny amount of Super Glue on the inside of the shaft can be used to lock the retaining rings in place. Be very careful not to get any of the Super Glue on the gears or outside of the shaft; you don't want to lock the gears in place.

PRINTING THE CAMERA MODEL

Unlike the gears, which are intended to be seen and turned, the camera model is really just serving as a water-resistant stand-in for the actual camera. It's perfectly fine to print this with the fastest print settings you're comfortable with. I used 0.4 mm layers and 5 percent infill to speed this print up. (See Figure 7-14.)

FIGURE 7-14: Printed camera model

LEATHERWORKING

There are a few required steps to put this project together, but many of the steps have multiple options, depending on the look you want to achieve. I'm using a black top hat, but other colors might dictate other color choices for the leatherwork. This project is primarily a skills description and source of inspiration, so feel free to make whatever choices suit your goals.

CLEAN UP AND TRIM

I frequently find that I like the look of the charred black edge left on the leather. Many leatherworkers would find this unacceptable. If you're uninspired by that look, you can easily trim the edge. (See Figure 7-15.)

Be careful; 4–5 oz leather can't sustain a very deep bevel. Make the shallowest cut you're comfortable with. If you dig too deep or find the edge too thin, you may be able to make repairs. You can use a very fine grit sanding sponge to even out a jagged edge, or even make another straight cut along the edge. Don't give up easily! You'll be surprised at how many projects you can save if you take a deep breath and consider alternatives.

The holes, especially the small ones for stitching, may have charred leather bits. It's much easier to clean all these out when the leather is fully dry. For small holes, it's best to use a hole punch to clean the holes out, rather than trying to sand or cut.

We'll be folding the leather in multiple places, and this will go much better if the fold lines are grooved on the flesh side. (See Figure 7-16.)

Mark the groove locations with a pencil line. Orient the point of the groover's V on the line, and place a straight edge along the groover's side. Be sure to check both ends of the line so that the straightedge is positioned to accurately guide the groover along the full length. Cut the channel shallower than half the thickness of the leather; you want to leave enough for strength in the bend. It's better to undercut and adjust than overcut, so start off with a small bite. (See Figure 7-17.)

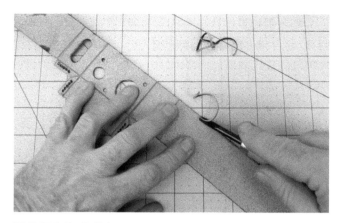

FIGURE 7-15: Edge trimming

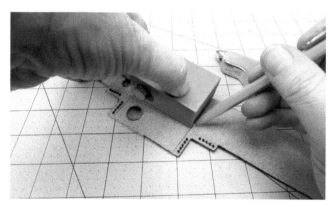

FIGURE 7-16: Groove locations for folding

FIGURE 7-17: Cutting the grooves

I decided to provide extra strength and a better visual look to the two parts of the holder where I would be sewing. Rather than just sewing the two edges together, I cut a small piece of leather that wrapped over the edges, and then stitched through all four layers. (See Figure 7-18.)

These pieces were ¾" × ⅝". Since I wanted them to make two folds to wrap over the leather edges, I needed them to have grooves to bend easily. Two grooves, ¼" apart, would do the job. It's hard to cut grooves on small pieces, so I made them on a larger piece first and then cut out my pieces. I nipped the corners off with a utility knife just to make them look a little nicer. (See Figure 7-19.)

COLOR AND FINISH

This project has both leather and plastic components. We'll paint both of these, but with different types of paint. A surprisingly wide array of paints will work on leather, but the paints that dry to a harder finish tend to crack over time when used on flexible leather.

Leather

I wanted a dark brown leather hatband with metallic patterning. My favorite leather dye, Canyon Tan, provided the base color, but I was concerned that it would be too dark of a base for the metallic paint. To provide a light background to paint over, I used Super Shene as a resist on the areas of the pattern I wanted to keep light. (See Figure 7-20.)

If you make the choice to not paint the pattern and just have the lighter, undyed leather show, you'll want to let the resist dry completely and use a second coat.

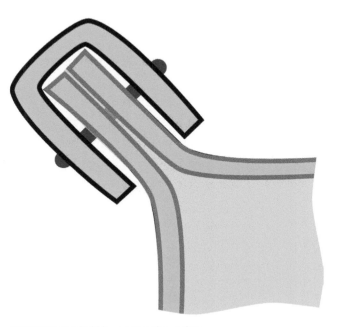

FIGURE 7-18: Wrapping the edge

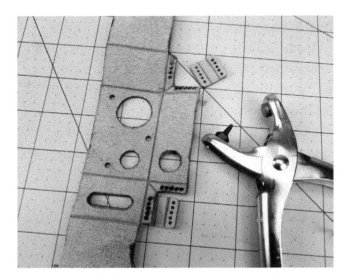

FIGURE 7-19: Prepping the edge-wrapping pieces

As noted, I'm a fan of using budget airbrushes for dyeing leather, but any method you're comfortable with will work. I used full-strength Canyon Tan (Eco-Flo #2600-06) to achieve my favorite color of brown. Full-strength dye goes on dark, but lightens as it dries. After my first coat dried completely, I gave the hatband a second coat. (See Figure 7-21.)

Don't forget to dye (and finish) the overwrap pieces we prepared in the previous section. You'll want them the same color as the hatband.

No matter how you apply your dye, you'll want to buff it before applying a finish. Use clean sheep's wool or a clean cloth. Buffing removes excess dried dye and gives the dye a gentle luster. If you don't buff, you'll get a slightly more matte finish, but you'll also impede the ability of subsequent products to adhere to the leather—not dramatically, but enough that it's worth the minute or two to buff the piece.

I wanted the pattern to pop, so I painted the light sections with a metallic acrylic paint formulated for use on fabric, paper, leather, wood, and polymer clay (Lumiere 565 Metallic Bronze). This product is intended to be able to cope with a degree of movement and flex without cracking. (See Figure 7-22.)

I applied the paint with two brushes. The first was a #2 brush to paint in the larger areas in the pattern. For the edges and tight spots, I used a very small #3/0 brush that I

FIGURE 7-20: Resisting parts of the pattern

FIGURE 7-21: Applying dye

FIGURE 7-22: Painting the pattern

usually use for miniature figure painting. (See Figure 7-23.)

If you find that painting inside the lines isn't something that comes easy, you can use the blade of a small utility knife to lightly scape away any stray paint, as long as you do so before it dries completely. You will notice that after a while it feels less like you're painting with a brush and more like you're painting with a stick. Every five minutes or so, clean the brush by wiping it on a paper towel, then dipping in water and repeating. It also helps to use the smaller brush to occasionally apply a tiny drop of water to your paint so it doesn't dry and form a skin. It also helps to put your paint (a small drop at a time) onto a damp paper towel instead of a dry surface. This makes dampening the tip of your brush a little easier, as well. For long sessions, you can use a misting bottle to help keep the paint liquid. Don't overdo it; too much water will thin the paint. That's not harmful; it just might change the way those sections look compared to the areas with full strength.

After the paint dries, you can apply a second coat. This is much easier than the first, in that you only have to hit the thin spots. Move the piece around to view at different angles, which will help you find areas that need more paint.

Once both coats are dry, or if you decided not to paint, apply your favorite finish to the grain (smooth) side of the leather. I like a glossier finish and generally

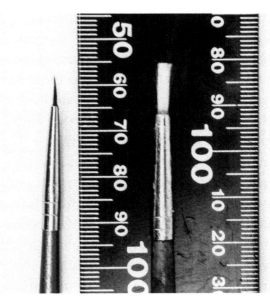

FIGURE 7-23: #2 and #3/0 brushes

use Resolene, but the Super Shene I used as a resist is also an excellent finish that I use on many projects. Leatherworkers develop preferences, some fiercely held and defended. I can't honestly say I find a dramatic difference between two products like Resolene and Super Shene, or other water-based gloss finishes I've used.

3D Printed Pieces

Your choice of paints will depend on the type of plastic you printed with. Not all paints will adhere to PLA successfully, but I've had great experiences using Testors® Enamel Paints. These are the little ¼ oz paints sold for painting models, and available in most craft and hobby stores. I've been using these paints for decades, so I have a soft spot in my heart for them. But painting with enamels is very different than painting with acrylics.

Enamel paints have a much stronger odor than acrylic paints. This is due to the mineral spirits that form the liquid base of the paint. As a result, you can't clean up with just water. You'll need to use paint thinner. The thinner sold by Testors is a propylene glycol monopropyl ether–based product with petroleum distillates and a few other things. It has a strong solvent odor and can cause respiratory problems in high concentrations. It is very important to have adequate ventilation. The threat is no worse than painting with most oil or enamel house paints, but breathing good air is important and not to be neglected. Clean your brush with appropriate thinner, or use disposable brushes.

Drying time is another issue with enamels. They dry to a very hard finish and look great, but take 48–72 hours to fully cure. If you handle them on the same day you paint, you will leave fingerprints and marks on the surface. Allow the painted pieces to dry fully before touching them.

I painted the gears and their holder, but avoided getting paint on the inside shaft of the gears and the outside shaft of the holder. These areas won't be seen and it's not worth making it harder for the gears to spin. I used a utility knife to pin pieces in place while applying paint. (See Figure 7-24.)

Let the parts dry for at least two hours before applying a second coat. I wanted the paint to hide some of the striation from the printing, so I applied three coats to the parts. (See Figure 7-25.)

SEWING THE CORNERS

Before starting to sew, take the overwrap pieces we prepared earlier, hold them up to the edge of the stitching holes in the hatband, and use a pencil to mark the locations of the holes. (See Figure 7-26.)

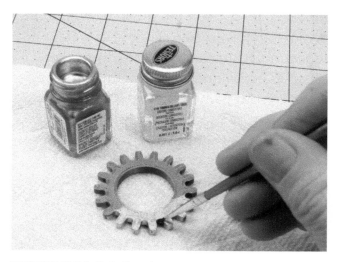

FIGURE 7-24: Painting the gears

FIGURE 7-25: Dry the parts for two hours between coats

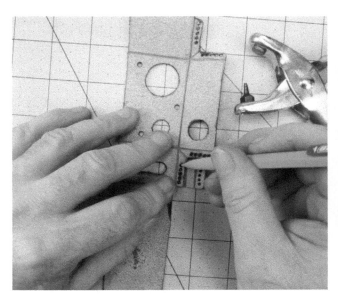

FIGURE 7-26: Marking the stitching holes

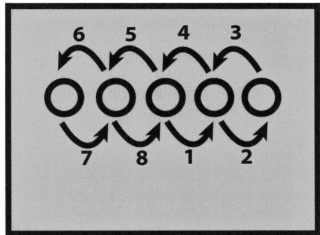

FIGURE 7-27: Stitching pattern

Use a sewing punch, stitching fork, or awl to punch holes through at the marked locations on each of the edge covers.

You won't need much thread, but use the two-needle stitching technique described in Chapter 2, "Fundamental Leatherworking Tasks and Tools," to sew through all four layers. Start in a center hole, stitch to one edge, stitch back to the other edge, and then back to the hole you started in. (See Figure 7-27.)

Trim the thread close to the holes on each side, and rub the thread on both sides with some beeswax.

RIVETING THE GEARS AND SETTING THE EYELETS

We're ready to attach the gears to the front of the holder. We'll use a product called Rapid Rivets. These rivets have a flat back and a snap-on cap. I'm not always confident they'll hold up under extreme wear, but they're perfect for mounting this gear plate to the hatband.

Push the rivet base through the hole in the face of the hatband camera holder and gear mounting plate. Place something solid with a hard surface underneath them and push the cap onto the shaft of the rivet. Position a concave rivet setter on top of the cap, and hammer firmly with a couple of good raps. The cap should grip snugly onto the shaft. Set all three rivets. (See Figure 7-28.)

The holes that were burned on the ends of each of the arms are for lacing the hatband tightly around the hat. To avoid stretching and tearing the leather, as well as to make the band look great, we'll place eyelets in the holes. Eyelets are a one-piece item. Push them through from the front of the hatband. Then, set the rounded front of the eyelet into the circular eyelet anvil. This allows you to set the backside without deforming the front. The setting tool has a convex dome that pushes the rim of the eyelet out until it splits and forms tines that grip the leather. Set all six eyelets. (See Figure 7-29.)

MOUNTING THE GEARS

The two gears should slide onto their respective shafts and be able to move one another. You can lightly sand the shaft or the inside of the gears if they're too tight. Once they're as you like them, push the retaining clips onto the front of each shaft. Check the gears again to make sure they still spin, and adjust if necessary. (See Figure 7-30.)

The retaining clips can't stay in place by themselves very well. You can take a toothpick with a tiny amount of Super Glue on it and apply the glue to the edge of the retaining clip on the inside of the shaft. This is easiest to do from the back of the holder.

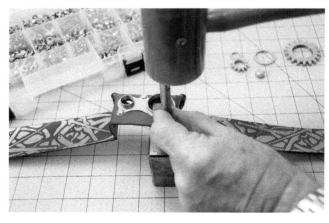

FIGURE 7-28: Setting the Rapid Rivets

FIGURE 7-29: Setting the eyelets

THE RESULTS

Lace up the hatband in the back. I used black leather lacing strip, but you could use any color of your choice. (See Figure 7-31.)

Slide the entire band up a few inches and slip the camera into the bottom of the holder. Slide the band, with the camera, back down to the base of the hat. Make sure that the little button that was printed is in front of the power button. (See Figure 7-32.)

FIGURE 7-30: Gears mounted

FIGURE 7-31: Lace up the back tightly.

FIGURE 7-32: The completed hatband

ENHANCEMENTS

There are an unlimited number of variations and enhancements possible with this design. Rather than gears, a 3D printed mechanical iris could form the front of the lens. (See Figure 7-33.)

Another approach to the design could involve providing a back and base for the camera part of the holder and adding a snap fit so that you could move the camera from hatband to wristband to penny farthing handlebars.

The patterns on the sides of the hatband also offer a range of choices. Additional gears could be mounted, and the adventurous Maker could even build a Rube Goldberg contraption that pushed the start and stop buttons of the camera. This approach is a platform capable of supporting whatever steam-, diesel-, or clockpunk ideas you're waiting to create!

FIGURE 7-33: Mechanical iris
IMAGE COURTESY OF WIKIMEDIA COMMONS

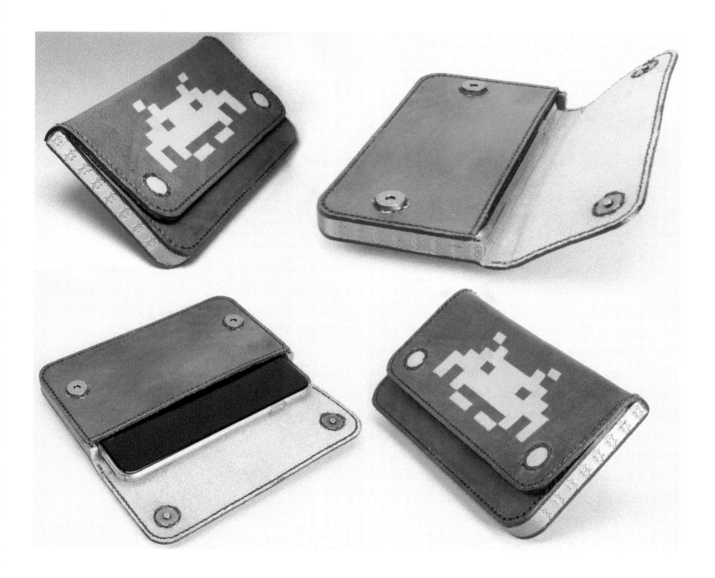

8-BIT CELL PHONE BELT CASE

It's an interesting phenomenon that, at a time when resolutions are increasing on every device, there is a nostalgia for the oldest form of computer graphics. These blocky images are often referred to as *8-bit*, in reference to the amount of memory space allocated to a pixel. The earliest video games did marvels with big square blocks, for which we willingly suspended our disbelief and saw aliens, frogs, apes, and aggressive plumbers.

DESIGN

I'll use the 8-bit aesthetic as our design motif in this cell phone belt case project. We'll use selective dyeing techniques to produce an 8-bit alien on the cover, and echo the image on the sides of the case. Of course, any images could be used in this approach. As with all the projects in this book, my goal is that you'll gain the skills to create your own designs rather than just re-create the specifics of these projects.

A cell phone belt case is a remarkably simple leatherworking project, or at least many approaches to the problem can be. Choosing how to make a case for a phone is an excellent example of how to gather requirements and approach product design. A cell phone case is best when it's a good fit for the phone, so the phone you choose makes a big difference in the design. I'm going to use an iPhone 6+ as the target phone, but that only starts the questions I need to ask.

I'm far too clumsy to carry a phone that doesn't have a rubberized shell attached. I need something to protect the phone if I drop it. Over the years, I've used a wide range of this class of casing, from hard cases that made the phone dramatically larger than stock to small, silicone cases that barely increase its size at all. I'm currently using one of the small cases, but understanding how the target user protects their phone is important to know, since it changes the dimensions. It's extremely frustrating to make a leather belt case for someone and discover their phone won't fit. (See Figure 8-1.)

There are other design decisions, as well. Horizontal versus vertical carry is a distinction that makes a big difference for many people. While some case designs work well in both orientations, many do not. Planning for the intended orientation is something that you must do from the start of the design. I currently carry a work phone and a personal phone, and have a vertical-mount, friction-fit case to hold them. (See Figure 8-2.)

One of the significant aspects of the orientation is the means of holding the case to the belt. Some commercial phone cases have clips that can turn 90 degrees to allow the case to hold in both positions. Many leatherworkers will add a standard belt loop like

FIGURE 8-1: Different protective phone cases

FIGURE 8-2: My dual-phone belt case

the one we used with the multi-tool holder. While I love the robustness and simplicity of the belt loop, the ability to remove the phone case without undoing my belt holds a strong appeal.

In the end, I chose a horizontal orientation with top access. I also chose a spring clip instead of a belt loop. They're slightly more work, but the ease of removal and the professional look made the decision for me.

Ultimately, to marry 3D printing and leatherworking in this case, I've made a printed frame that wraps around three sides of the phone. The front and back panels will be leather, with the back panel folding over the top and onto the front, where I will dye the 8-bit alien pattern.

I wanted to try something I'd never done before with this design and incorporate the 3D printed frame into the stitching of the leatherwork. There are many ways to attach a plastic frame to the leather pieces; rivets would have been an easy choice. Instead I included channels through the frame that matched the stitching holes on the leather panels. The stitching will go through the leather, then the frame, and then the other leather panel. This provides an even attachment around the frame and takes advantage of the attractiveness of leather stitching. (See Figure 8-3.)

It's important to be able to answer calls quickly, so I'm using magnetic bag clasps to hold the front cover on. This allows easy opening and a very forgiving closure that doesn't take a lot of fussy activity.

TAKING MEASUREMENTS

Phones are pretty easy to measure—we'll need a height, width, and depth. Pay attention to protrusions such as power or volume buttons. Unless your design has special accommodations for them, you should treat them as the outside edge of the dimensions. Curved edges and corners can be ignored or included in the design. I decided not to worry about the curves on the edges, but

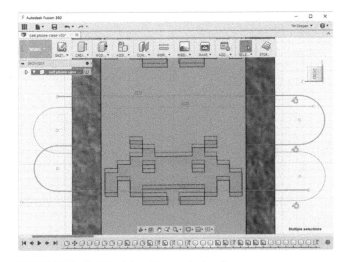

FIGURE 8-3: Stitching through the frame

DESIGN **131**

wanted to fit my case to the curves, referred to in CAD as *fillets*, on the corners. (See Figure 8-4.)

Cases can be designed for a snug or loose fit, but if they're too loose, the phone will rattle around uncomfortably. I added a half a millimeter around the edges of the phone on the three sides where it will fit into the case frame. I made this change to the frame model, but I did not expand the model of the phone. This allows me to view the gap when looking at the CAD models. (See Figure 8-5.)

THE 3D PRINTED FRAME

The frame seemed easy; a simple U-shaped structure the height of the phone with some holes in it. I don't like a big bulky frame, so I didn't want the frame any thicker than it had to be. I started off by simply adding a 6 mm wide border around the phone. To get it

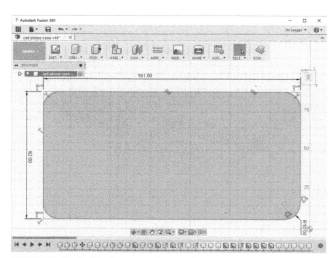

FIGURE 8-4: My target phone dimensions

a little less snug, but still keep it tight, I took off 0.25 mm all the way around. I cut off one long edge of the frame to make it U-shaped. A few fillets to take off the rough edges and I had my basic shape. I sketched a center line around the frame to provide a guide for the stitching holes. (See Figure 8-6.)

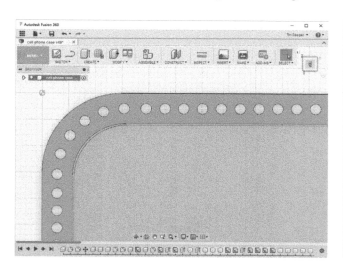

FIGURE 8-5: Gap around the phone

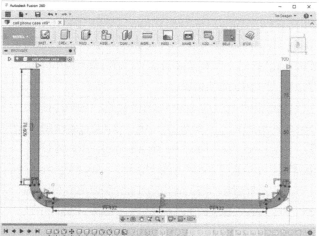

FIGURE 8-6: Midline defined on the frame

I discovered, after some experimenting, that the holes had to be larger than I would have made them if I were using just leather. Obvious in retrospect, the width of the needle, which stretches leather, can't do so with plastic. I determined that I needed 2 mm holes. The distribution of the holes had to include the rounded corners, and I ended up using the Circular Pattern and Rectangular Pattern features of Fusion 360 to distribute the holes. As long as you leave enough leather between the holes to keep the stitches from tearing—at the very least 1.25 mm for 4–5 oz leather—the distribution isn't critical.

I also wanted to use the frame to echo the 8-bit alien motif, so I recessed the pattern multiple times down each side. This was only 0.3 mm deep, but that's enough to see it easily without creating aggressive overhangs for the printer to deal with. (See Figure 8-7.)

THE BASIC LEATHER PATTERN

The leather pieces for this project are relatively uncomplicated. There is a rectangular piece sewn to the frame's front. Another piece is sewn to the back that is long enough to fold over the front and serve as a cover and flap. Then there is a smaller rectangular piece that will be sewn to the back flap to hold the belt clip. (See Figure 8-8.)

Front

The front piece is a simple cut: sharp corners at the top and fillets to match the frame at the bottom. Of course, the stitching holes have to match the ones on the frame, but I

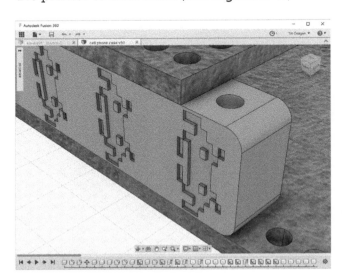

FIGURE 8-7: Alien pattern insets

FIGURE 8-8: The three leather sections

was careful to extend the top hole on each side just far enough that it would be above the edge of the leather. (See Figure 8-9.)

The reason for this additional hole is to provide for an extra stitch that extends over the edge of the leather. The leather's edge is where the most wear and stress will occur, and the last stitch is the point where it is the greatest. The extra stitch binds the edge hard against the frame, and is considerably stronger than allowing the leather to pull down against a stitch.

The other area where this piece will experience extra stress is across the edge where the phone will be slipped in. Leather stretches with wear unless you provide a support. A row of stitches across the edge doesn't attach the leather to anything, but it does significantly enhance the leather's ability to handle usage over time. (See Figure 8-10.)

Back Flap

The back flap matches the front piece's stitching holes around the frame, and has its own set of reinforcement stitches around the edge. It also has stitching holes to attach the belt clip cover. (See Figure 8-11.)

The length of the flap is a little difficult to calculate, since the way the leather folds over the top will impact how much extends down the front. To get an approximate idea, I created a "folded" model of the back flap with the edge where I wanted it, and measured the outside edges to get the total length. We'll improve the fold quality with grooves, so measuring the inside is less effective. Of course, leather stretches and has lovely organic qualities, so leaving room for surprises is always a smart idea. (See Figure 8-12.)

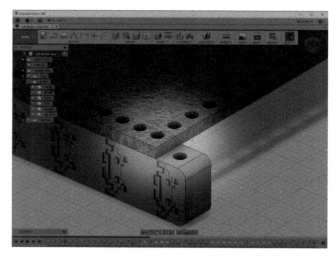

FIGURE 8-9: Extra stitching hole

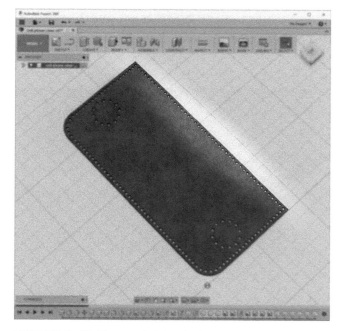

FIGURE 8-10: Front stitching holes

Clip Cover

Since the design decision was to use a spring steel belt clip, care has to be taken not to have any metal in the case that might scratch the phone. Phones are far more scratch-resistant than they used to be, but the practice of considering what will be up against your item in its new case is important for any number of projects.

The clip I've chosen has four rivet mounts. Rather than riveting these to the back flap, which would leave a rivet base or head inside the case, I'll rivet it to a smaller clip cover. This cover will be stitched to the back flap on top of the clip to hold the clip in place. You must allow room for the clip to emerge, so cut a small indentation at the top to allow room for it to exit. If I'd made the clip cover bigger, with the stitching holes farther out, I could have used a slot, but I wanted a small cover. So to reinforce the edge where the clip will emerge and be pulling, I added an extra hole on the back flap, where I can do the same reinforcement stitch I did on the front piece and the frame. (See Figure 8-13.)

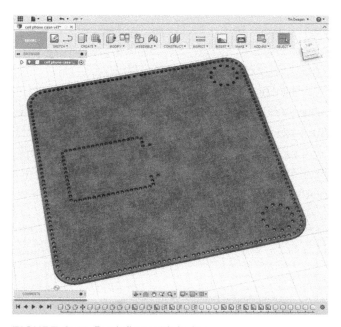

FIGURE 8-11: Back flap with holes

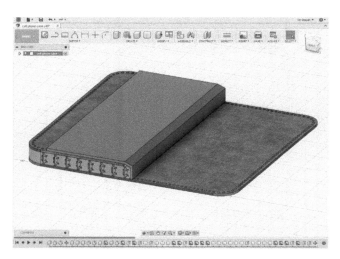

FIGURE 8-12: Folded back flap

FIGURE 8-13: Clip cover

DESIGN **135**

DIGITAL FABRICATION

There are three digital fabrication techniques involved in this project. We'll use a CNC with a 1/16" ball mill to cut the leather, a 3D printer to print the frame, and a stencil cutter to prepare the vinyl stencil for the dye work. The outline cuts for the leather could have been done with a drag knife, and all the leatherwork could have been completed with a laser cutter, but the CNC router is a tool we haven't used in previous projects and is loaded with potential. I'm using Vectric VCarve Pro for generating toolpaths, but Easel, a web-based CAD/CAM tool from Inventables (*https://www.inventables.com/technologies/easel*) is a great free alternative.

Any 3D printer with sufficient build volume should be able to print the frame. Smaller printers may need to orient wide items like this 45° to fit. Print resolution is a matter of preference and capability.

The stencil cutter I'm using is an older model Silhouette SD, but any stencil cutter will work. The stencil could be easily cut manually from vinyl, tape, or another adhesive-backed material.

CNC WORK

Arguably, all the cuts could be included in a single G-code file if the workpiece is large enough to lay everything out. For this project, this would require a piece of leather at least 280 mm tall by 250 mm wide. I usually prefer to break the files up into individual part files. Then, if something goes wrong, I only need to rework one piece. It also allows me to make more efficient use of my leather stock, since I don't have to find a larger, homogeneous piece. But the single cut file is a great way to knock this out quickly. (See Figure 8-14.)

I did choose to combine all the cuts, both holes and outline for each piece, into a single file rather than running them separately. This was easy, because I didn't have a tool change operation between the two types of cuts. It's not impossible to set up G-code to prompt

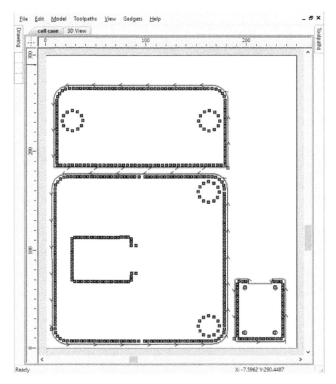

FIGURE 8-14: Pieces laid out for cutting

for and allow a tool change, but it's more complex and provides opportunities for trouble. On my CNC, the cutting tool is a DeWalt DWP 611 trim router. This is a common tool used on desktop CNC machines. Changing tools involves pressing a spindle lock button on the side of the unit while loosening the collet nut with a wrench.

Unless you've gone to great lengths and are very careful, it is extremely difficult to remove one tool and put in another so that the tip of the tool is at the exact same height. This is a big deal, since the machine only knows where to move the tool based on where you told it the Z axis zero location is. If the new tool is 3 mm higher than the old tool, the CNC will think it's cutting at the same location, but it isn't. Common practice is to re-zero the Z axis on every tool change.

There are other challenges, as well. If the nut is tight or something slips, it can be all too easy to inadvertently move the tool in either the X or Y direction. At that point, you cannot continue the cut without rehoming the unit. All of this argues in favor of breaking up your G-code into multiple parts.

As noted, there are two types of actions on each of the pieces. The first is cutting out the holes, and the second is cutting out the outline of the piece itself. These can occur at the same work depth and with the same 1/16" bit I described. (See Figure 8-15.)

The significant difference is that the holes need to be cut from the inside and the outline from the outside. The reason for this is

FIGURE 8-15: 1/16" ball end mill

the need to try and stay true to the dimensions of the target piece. Even a 1/16" bit cuts out *kerf*, the part of a board or workpiece that is lost to the width of the cutting tool. The cuts always need to occur on areas that are waste. If the outline is cut from the inside, the piece will be 1/16" short on all sides.

It's also a very good idea to always take advantage of the preview or visualization tools available in the various parts of the toolchain. Fusion 360 felt confident that the model would look great. Checking to see if the toolpath software also thinks things will work can save a great deal of grief. (See Figure 8-16.)

PREPPING THE WORK PIECES

Unlike the laser, whose pressure takes advanced scientific tools to measure, the CNC router is going to drive a high-speed bit into the leather with significant force. If the workpiece isn't firmly held down, the bit will push it around the work table.

Leather isn't a good candidate for holding with clamps. If you clamp down the edges, but cut patterns out of the middle, it's possible that unsupported areas will lift or move under pressure as the area around them is cut out. Commercial leatherworking companies frequently use a vacuum table to hold leather in place during cutting, but a sufficiently strong vacuum with enough points of contact (holes in the board) is beyond many home shops.

Double-sided tape is a common solution, but it has its own problems. Using strips of tape can lead to an uneven surface under the leather, and the entire area that will be worked really needs to have an even surface under it. Double-sided tape is also surprisingly expensive when you start using it regularly to cover square feet of surface area. And lastly, except for very expensive or very thick tape, it tends to have fairly lame holding power.

Luckily, there is an inexpensive alternative that is easy to use and works great. Ben Crowe, a master luthier at *https://crimsonguitars.com/*, has worked out a method that has become very popular. He lays down strips of wide masking tape on the work surface and on the workpiece. He then covers the tape on the work surface with cyanoacrylate (Super Glue) and places the tape-covered work piece on top of it. The method can also be used with a cyanoacrylate accelerant spray to get a very fast grip.

While this method is a huge win over double-sided tape, I've taken it a step further. I don't like lining up strips of tape, and even masking tape gets more expensive as it gets wider. I've also made huge messes with Super Glue on a number of occasions. Instead, I buy rolls of 12" wide adhesive shelf-lining material. This is typically made of vinyl, backed with a non-permanent adhesive. I cut two pieces: one for the table and one for the workpiece. If necessary, I can place parallel strips of shelf paper for larger pieces, but I also have rolls of 24"- and 36"-wide material if needed. It's important to not get air bubbles between the leather and the liner, but the shelf liner is good about peeling partway off to let the bubbles out, then smoothing back down.

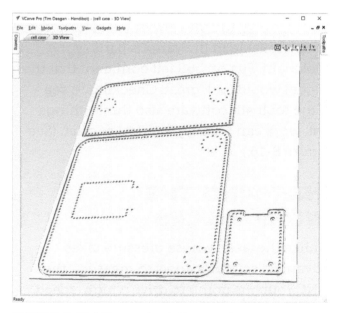

FIGURE 8-16: VCarve simulation preview

Worst case, you can always prick a bubble with a pin.

Instead of cyanoacrylate, I use aerosol adhesive, usually Aleene's® Repositionable Tacky Spray, though stronger sprays are available. (See Figure 8-17.)

I mask the area around the section on the table and then give the section a quick spray. (See Figure 8-18.)

I can then position the leather workpiece on top. If I get it misaligned, I can remove it and reposition if necessary. Once it's in place, I give it an even rub to smooth it out. The holding power is outstanding, and removal is easy. The shelf liner peels away from the leather without any problems.

When milling with the bit I'm using, the results are best if the pattern is reversed (or symmetrical) so that you can place the leather face down and cut from the backside. The shelf liner is easier to apply and remove from the skin side, so this works out well. (See Figure 8-19.)

CUTTING OUT THE PIECES

The 4–5 oz leather I used was approximately 1.5 mm thick. I wanted to cut all the way through, so I set my cutting depth for 1.9 mm. I gave myself 0.4 mm margin of error because I was using an unsurfaced spoil board.

FIGURE 8-17: Tacky Spray and shelf liner as custom double-sided tape

FIGURE 8-18: Masking and spraying adhesive

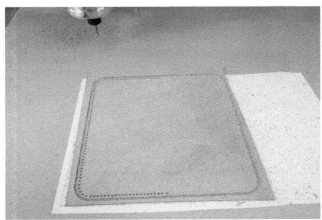

FIGURE 8-19: Leather held for milling

Surfacing a Work Table or Spoil Board

The table of a CNC can be metal or wood. It might have T-slots or inset threads for mounting, or nothing at all on some DIY CNC machines. It is a common practice to try and preserve the base surface of the CNC from cuts and accidents by placing a sacrificial *spoil board* underneath the workpiece. MDF is a useful spoil board material due to its very consistent properties, but even small pieces may have some degree of bend to them.

The CNC, from the point of view of the software, operates in a completely level environment. The work surface is expected to be the same height at all points. This is a theoretical model, and as the saying goes, "The difference between theory and practice is that in theory there is no difference." Ultimately, the real world pokes its nose into the problem.

If the spoil board, or even the CNC table, isn't perfectly flat, then the cutting tool will make cuts of varying depths as it passes across the surface. (See Figure 8-20.)

The solution to problems like this is to "surface" the work surface, either the spoil board, work table, or both. Special tools, called *surfacing bits* or *fly cutters*, are used to do this, since using a ¼" end mill would take forever and result in a marginal finish. Surfacing bits have a wider cutting area, the width of which depends on the power and size of the tool used as a spindle. For the DeWalt on the Inventables X-Carve I use, a 1 ½" diameter surfacing bit is about the largest practical bit to use with the ¼" collet on the router. (See Figure 8-21.)

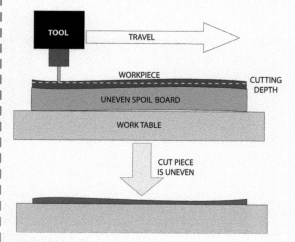

FIGURE 8-20: Effects of an uneven surface

FIGURE 8-21: Surfacing bit

> By setting the height of the bit at, or just below, the lowest point on the surface, the CNC can be programmed to make a series of passes that cover the entire surface, cutting it level. If the amount to be cut is deeper than the effective pass depth of the surfacing bit, it may require multiple passes. The Amana RC-2248 I use suggests a maximum cutting depth per pass of ¼", but in practice I'd keep it under ⅟₁₆" in most cases.

CUTTING THE PATTERN STENCIL

The pattern I've chosen is a classic 8-bit alien from Wikimedia.org. (See Figure 8-22.)

FIGURE 8-22: 8-bit alien

This is simple enough that it could be manually traced and painted with resist. But since we're exploring digital fabrication techniques, the stencil cutter is a great solution.

Stencil cutters are one segment of the digital fabrication market. They, and computerized embroidery machines, are marketed to a very different segment than the traditional Maker and DIY communities. They are aimed at the crafting and fabric arts communities. This is a shame in many ways, since these tools are extremely useful for a wide array of tasks. But their toolchains have to work for users who think they're a non-technical audience (despite the fact that many crafters and fabric arts people do the equivalent of some very complex CAD work!).

The Silhouette model that I'm using is a generation behind the current products, though they really only differ in the small ways. The newer models tend to have a wider cutting frame and are moving towards an online toolchain, much like the CAD/CAM Easel tool from Inventables that I mentioned earlier.

Commercial vinyl cutters are functionally identical, though they are designed for heavier use and wider material. A drag knife in a CNC machine is also pretty much the same thing, just a bigger knife. (See Figure 8-23.)

I imported the image into Fusion 360 and traced it on a sketch layer so I could turn it into a DXF. I could have used Inkscape or Illustrator for this, but Fusion made it easy to test the pattern's position on the model of the case, which helped judge the appropriate size. (See Figure 8-24.)

Once the size and additional elements, like the circles I decided to put inside the magnetic snap stitching, were laid out, I exported the DXF and imported it into the Silhouette Studio program provided for use with my cutter. I liked the image enough that I decided to cut three, both for backups and rogue stencil graffiti. (See Figure 8-25.)

The selected lines for cutting are highlighted in red in the image. The tool allows for modifications, but in this case, the only important one was scale. The DXF initially imported to fill the page size, but I took measurements from Fusion 360 and used them to rescale to appropriate size. Selection of adhesive-backed vinyl and the appropriate knife were the last two configuration steps. I did reduce the thickness after a test cut, since the cutter should ideally cut through the vinyl without cutting through the backing paper.

Once the stencil is cut, we need to prepare it for transfer. This can be done as a positive or negative image, depending on which

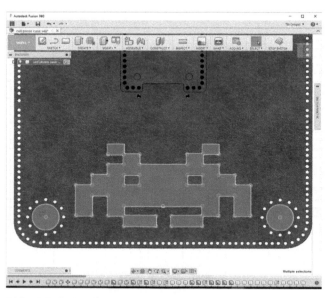

FIGURE 8-24: Overlay of the pattern on the model

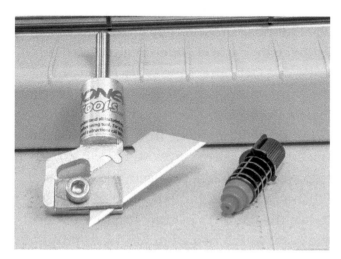

FIGURE 8-23: CNC drag knife vs. stencil cutter knife

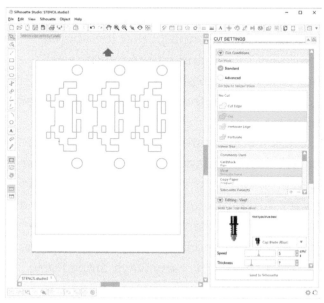

FIGURE 8-25: Alien in Silhouette Studio

parts are left and which removed. I want to dye most of the cell case a warm brown, but leave the alien pattern in a lighter shade. I'll retain the positive pieces of the stencil, the ones that look like the alien, as opposed to removing the alien and leaving the negative surround. (See Figure 8-26.)

An alternate approach is to use the negative stencil as a pattern to spray resist onto the leather. Nearly every attempt I've made to use this process has ended up with resist leaking under the edges and causing the pattern to blur. (See Figure 8-27.)

This can also happen spraying dye onto a positive stencil. The only way to avoid it is to go very slowly, being extremely careful not to oversaturate the leather with dye, which will cause the dye to soak into the leather enough that it seeps under the edges of the stencil. I would normally spray dye onto leather in three to four coats. Using the vinyl stencil approach, it will take eight to 12 much lighter coats. I'll describe the process more in a later section.

With the unwanted parts of the stencil picked away, 8" or wider transfer paper is adhered to the front of the stencil. This stencil is very simple; many will have dozens or hundreds of tiny complex pieces to pick away. You can compare our stencil to one of Ganesha I've been working on. (See Figure 8-28.)

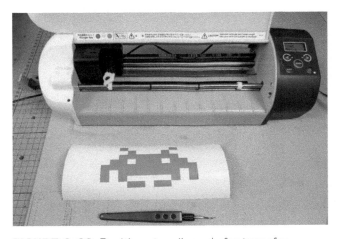

FIGURE 8-26: Positive stencil ready for transfer

FIGURE 8-27: Resist that leaked around edges of vinyl stencil

FIGURE 8-28: Complex Ganesha vs. simple alien

DIGITAL FABRICATION

The transfer paper is smoothed onto the stencil and will lift it off the backing paper. You can then smooth it onto the target, and ideally the stencil will stick harder to the target surface than the transfer paper and you'll be able to peel the transfer away. On occasion this doesn't work, since the vinyl doesn't stick to everything well, but veg tan leather is pretty consistent in its ability to hold the vinyl stencils. (See Figure 8-29.)

Make sure that all the edges of the vinyl are smoothed tight against the leather. To achieve a crisp edge on the dye work, we can't let any dye leak or seep under the vinyl.

PRINTING THE FRAME

The frame model produced in Fusion 360 was exported as an STL file. I'm printing with a Prusa i3 MK2 printer, which comes with a copy of Slic3r, which has presets for the Prusa. (See Figure 8-30.)

The infill hardly matters, since there really isn't enough open space inside the model to matter. I used 0.15 mm layer height, which is the optimal setting for the Prusa. Pushing the resolution higher seemed inappropriate for the 8-bit motif.

The model is so narrow, and so much of the interior is made up of the stitching shafts, that the top and bottom layers look the same as all the internal layers. (See Figure 8-31.)

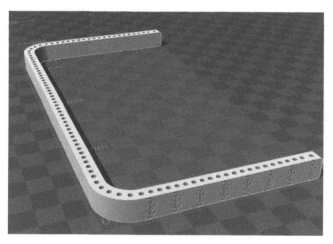

FIGURE 8-30: Frame model

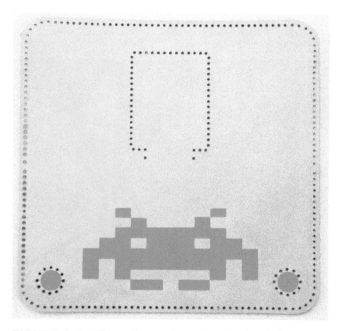

FIGURE 8-29: Stencil transferred to the back flap

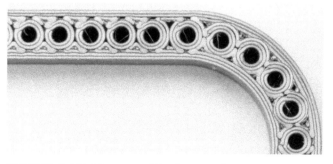

FIGURE 8-31: Closeup of the bottom layer

For some reason, I had a very difficult time printing this model. If something like this gives you trouble, take heart! Try and try again, slightly changing parameters until you get what you want. (See Figure 8-32.)

The color you print with is a personal choice. I considered black and other options, but ended up with gray as a neutral selection. It will be set off by the black edge paint I'm using in the coloring step. You will want to choose something that works best with the dye color you choose.

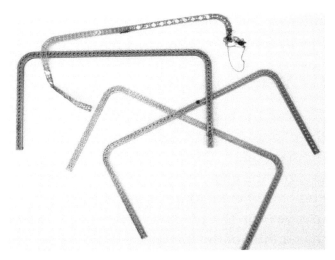

FIGURE 8-32: Many fails

LEATHERWORKING

This project doesn't have a lot of complicated leatherworking. Hopefully, by now you're itching to start stitching. Unfortunately, we need to do the dye work before we start stitching so that we can get a consistent look and finish.

CLEAN UP AND TRIM

The leather coming off the CNC has a nice edge and clean cut on the front, but there are many extraneous fibers along the back edge. You can trim these with embroidery scissors or a utility knife, but I prefer to use an emery board and file them off. (See Figure 8-33.)

If you use this approach, be careful not to file down the crisp edge on the front. Many leatherworkers will choose to bevel the edges. That's a personal choice, and it can look great; it can even improve wear to a slight degree, since it makes the edge less likely to catch on things and tug against the

FIGURE 8-33: Filing off unwanted fibers

stitches. I preferred to keep the boxy feel that the hard edge gave me.

PREPPING THE FOLDS

To assist the back flap in easily folding over the phone, it helps considerably to add two grooves: one at the line where the back flap touches the edge of the frame in the back, and one where it touches in the front. The grooves don't need to be very deep; no more than a third of the thickness of the leather. You can always wet the leather along the groove and bend it to improve the fold line. You could just mold the leather instead of grooving, but the grooves provide a more natural drape to the leather and keep the front flap from trying to spring up. (See Figure 8-34.)

COLOR AND FINISH

Because we're using the stencil dyeing approach in this project, this step is more tedious than usual. We'll need to do many light passes with the dye. I suppose it's possible to dye with the stencil resist method without using an airbrush, but it would be very difficult. The airbrush allows for even distribution, and can be applied in very small bursts.

Triple check the adherence of the stencil each time you make another pass with the airbrush. I'm using my old favorite, Eco-Flo Canyon Tan, and I'd like to get it pretty close to at least three-quarter strength

FIGURE 8-34: Grooving the back flap.

when completed. The problem is that I can't really let the leather get wet with dye. It has to be barely misted on each pass. You can't spray enough to even dampen the leather. I find that no more than two short bursts on small areas, or one quick, even pass, is all the leather can take before it starts to get wet.

The vinyl can serve as a guide. As you spray the piece, the dye can't soak into the vinyl at all, so it beads. As soon as you see beads, you know to stop. If the dye pools on the vinyl, it will puddle over the edge and seep under. Try to stop before you see pooling, but if you do, use a clean, dry paper towel to blot the piece and soak up extra dye. (See Figure 8-35.)

The leather needs to fully dry between applications of the dye. Once again, the vinyl is a useful indicator. Don't spray on another application of dye until the beaded dye on the vinyl has fully dried.

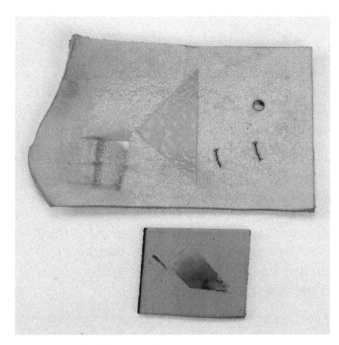

FIGURE 8-35: Dye beading on vinyl

This technique has fantastic potential, but is finicky. I highly advise practicing with some snipped pieces of vinyl and scrap leather before committing your actual project parts. Worst case, you can cut another one. Also, if you're applying dye instead of resist, you can go ahead and dye the entire piece completely and hide your sins.

Once you've completed the dye passes and the leather is the target shade, remove the vinyl and buff it with a clean cloth to remove the excess dye. Use your preferred finish to give the pieces a protective coat. I'm using Eco-Flo Super Shene on this project, but frequently use Fiebing's Acrylic Resolene. Both are acrylic finishes that provide a degree of water-resistance and add gloss and durability to the leather. (See Figure 8-36.)

ATTACHING THE SPRING CLIP

The spring clip is going to be held in place by both four rivets and the stitching of the clip cover. Since we have the additional strength that stitching provides, we don't have to use the strongest possible rivets. I'll use four Rapid Rivets to hold the clip to the cover since they're easy to set and are reasonably strong.

We used Rapid Rivets on the hat band project, since they're easy to set and have the advantage of having a flat plate on one side (they also come with domes on both sides). Slide the clip onto the cover so that the plate is on the inside. Push four of the small rivets through from the back and set the heads on the front side. (See Figure 8-37.)

FIGURE 8-36: Leather pieces dyed and finished

LEATHERWORKING **147**

Choosing the Right Rivet

Choosing the right rivet is usually easy, but occasionally presents interesting problems. For strength, a copper or brass traditional peened rivet is outstanding. These have a small learning curve and frustrate many users. They would have been my choice in this case, since they have a flat place that would have worked well on the outside of the leather. The problem is that I would have been using a strong mechanical link between two dissimilar metals on an item intended to be worn on the body.

This isn't a dangerous option, but you may have known people, like I do, whose sweat makes things around them rust. A guitar player in a band I was in had to replace all the screws in his guitar every year or so because they corroded. Attached dissimilar metals are a potential source of galvanic corrosion. Copper attached to stainless or iron-bearing steel is a higher-than-normal risk. Throw in an electrolyte like sweat, and the potential for gnarly stains on the pretty leather is high enough to try something else.

Tubular rivets are appealing, given their ease of setting and reasonable strength, but they have a raised rim on the flat side and tines on the gripping side. The raised rim could be frustrating to slide over a belt (I hate it when I struggle with clipping things on or off my belt). Additionally, the tines wouldn't work unless they're on the leather side. They need to bite into leather to achieve their full strength. Tines between the belt and case would scratch and shred the belt, which is not a good feature for a belt case.

FIGURE 8-37: Rivets set on the clip cover

Once the spring clip is firmly attached, stitch the clip cover to the outside of the back flap. Start at the corner, stitch back toward the center, and add three stitches up to the extra hole in the back flap. Back-stitch over the stitches to the corner and continue all the way around, ending the stitching pattern in reverse. Ultimately, you should finish stitching in the top-right corner. (See Figures 8-38 and 8-39.)

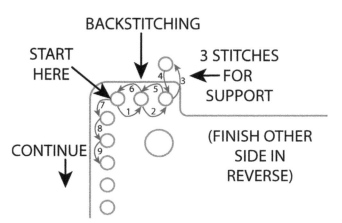

FIGURE 8-38: Stitching hole order

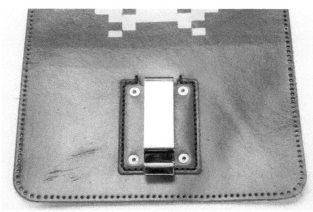

FIGURE 8-39: Stitched clip cover

SEWING THE CLOSURES

Most magnet snaps have tabs that pass through the material and are folded against a support plate. Without another cover sewn over them, these would have been a scratch risk for the phone. Sew-in magnetic snaps are also available in a small range of sizes and configurations. Many sew-in models are designed for hidden use, and intended to be sewn into clothing or drapery and invisible to the user. (See Figure 8-40.)

FIGURE 8-40: Push-through magnetic snaps

The snaps I'm using have four rings attached as stitching points. These snaps are available in various sizes and colors online. I'm using 15 mm bronze snaps purchased on Amazon. To attach them, I created a circle of stitching holes, which gives us one hole under each ring and one on either side. This allows a continuous circle of stitches to hold the snap in place. (See Figure 8-41.)

It's not critical where you start stitching, but you'll want to backstitch for at least three holes. For an even-looking stitch

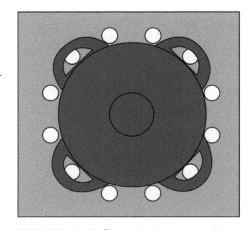

FIGURE 8-41: Stitch hole pattern for snaps

pattern and to add strength, I backstitch all the way around the ring. The needle will only fit through in one orientation, and barely at that. You may need to help the process using a pair of pliers; if so, go slowly. (See Figure 8-42.)

SEWING THE COVERS

Stitching the front cover and back flap to the frame is unique in a couple of different ways, due to the width of the frame. As in all my leatherworking projects, this is a two-needle saddle stitch covered in previous chapters. The challenge is to avoid having a stitch travel across the top of the frame when switching from the frame stitches to the supporting stitches on the edge of the front cover and the rim of the flap.

The easiest answer is to treat the stitching as three separate activities: stitching to the frame; stitching the front, top edge; and stitching the rim of the flap. With this approach, the stitching between these three areas doesn't actually connect. It is extremely difficult (unless you rework the sizing of specific holes) to backstitch in the frame section. To make this easier, start the stitching at the top of the frame and work toward the bottom middle. Do this separately from each side. With care, you can overstitch one hole at the bottom. If you're concerned, put a tiny drop of cyanoacrylate (Super Glue) onto that last overstitch. One of the wonderful things about

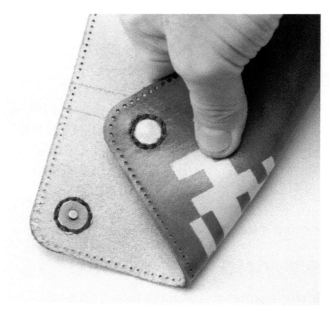

FIGURE 8-42: Stitched snap

hand-stitched leather is the ability to snip it all out and redo it if you don't like it later. (See Figure 8-43.)

To avoid the appearance of gaps, you can stitch the edge supports before you stitch the covers to the frame. Extend your stitching so that it includes the last hole: the one on each side that would be part of the frame stitching. This will make it difficult to get the stitches all the way through when you stitch the cover, so you may need to slightly enlarge the holes that are doing double duty.

Stitch the covers to the frame. It may require the assistance of some needle-nose pliers if the fit is tight. Go gently. If necessary, use a 1/16" drill bit to ream out the passage until things fit easier. (See Figure 8-44.)

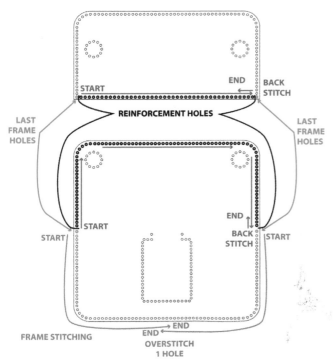

FIGURE 8-43: Stitch pattern for the cover

FIGURE 8-44: Stitching complete

LEATHERWORKING **151**

THE RESULTS

This design is able to support an unlimited range of motifs or decorations. It's a lightweight, but strong, structure that fits securely on the belt. The basic frame and cover shapes could be in a variety of patterns, not just rectangular. You could make custom-fit cases for all kinds of devices using this approach. (See Figures 8-45, 8-46, and 8-47.)

FIGURE 8-45: Cell case front

FIGURE 8-46: Cell case inside

FIGURE 8-47: Cell case back

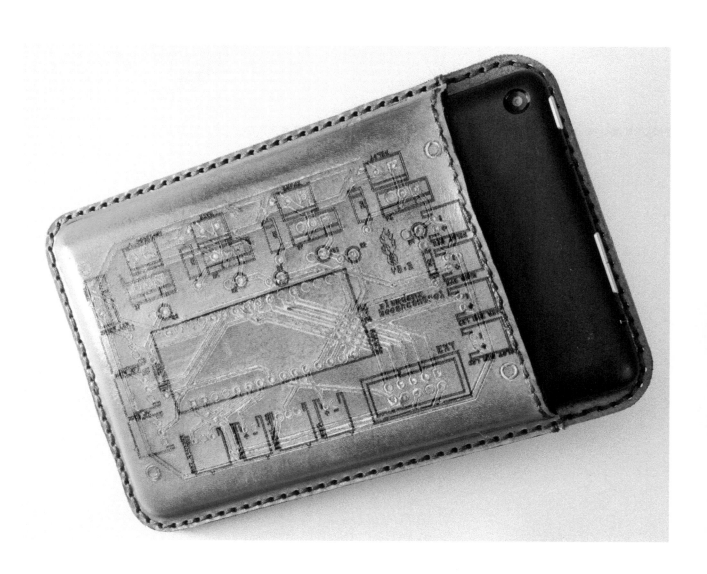

PCB TABLET SLEEVE

Tablets, large-screen phones, readers, and phablets have become ubiquitous artifacts in our daily lives. They are a science fiction staple turned reality that we now take for granted. Your life may be different, but damage to these devices seems to be a regular occurrence in my household. Budget manufacturing techniques may have a great deal to do with this, but wear and tear is a significant contributor. (See Figure 9-1.)

FIGURE 9-1: Broken screen phablet

DESIGN

All hail the conquering leatherworker! She will save these precious items by swaddling them in beautiful veg tan–preserved cow skin. Techniques dating back to the dawn of tool-making serving to enhance the harbingers of the coming singularity—what could be more satisfying?

With that lofty goal in mind, this project seeks to provide an easy-access, mid-sized tablet sleeve. Nothing about the project is specific to any particular set of dimensions. I am using an Amazon Fire tablet (7.5" × 4.5" × 0.4", or 191 mm × 115 mm × 10.6 mm). You can scale the dimension appropriately for your project, but confidence in starting from scratch with your own measurements is a key takeaway in this chapter and the book as a whole.

STRUCTURE

The sleeve is very basic. It's an open-top, slip-in design. Adding a flap or strap to hold the tablet in place, or including a belt clip or loop, is trivial. I find that I typically want to throw my tablet in my carry bag, so I'm primarily looking for something fast and easy to use that provides basic bumper protection. (See Figure 9-2.)

The thickness of the leather isn't really an important design consideration, so I'll continue with using the 4–5 oz leather we've been using in the projects so far. It ranges from 1.5–2.0 mm and tends to be priced economically.

The sleeve provides opportunities for decoration, primarily as a canvas on the front and back. I'm averse to adding conchos, since I don't want any metal on the inside of the sleeve. As we saw with the belt clip on the cell phone case project, there are ways to avoid metal inside, but I want a simple sleeve and will stick to non-metal decoration.

The design consists of two parts: the back, which remains flat, and the top, which is molded to the shape of the tablet. The top isn't as long as the back so that the tablet can be easily removed. The two layers will be stitched together, and both the back and front will have their remaining edges stitched for reinforcement.

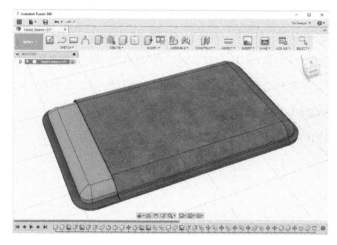

FIGURE 9-2: Sleeve design

Using the CNC to cut our stitching holes is more difficult in this project. The bottom, which is flat, is easy. But the top must go through a stretching process when molded, so precise prediction of where to place the holes isn't feasible. Leather isn't predictable enough to get that accurate consistently. The solution is to mold the top first, then cut the holes in the top and bottom at the same time.

This requires using a bit with a cutting length deeper than the height of two layers of leather stacked together. Instead of the 1/16" ball end mill I've been using, I'll switch to a 2-flute straight grooving end mill with a 2 mm cut diameter. (See Figure 9-3.)

We'll make a 3D printed model of the tablet that will be used both to mold the top and to remain in place for structural reinforcement when we mill the top and bottom layers of the leather at the same time.

ART

The motif for this project is a bit of an homage to my other book, *Make: Fire, The Art and Science of Working with Propane* (Maker Media, Inc; April 15, 2016). One of the projects in that book is a *boosh controller*, an Arduino-based control board for propane flame cannons. The printed circuit board (PCB) for the boosh controller is, like all PCB layouts, visually appealing to me. Since so much of modern society relies on electronics, I'm using the PCB design as the image for the sleeve. (See Figure 9-4.)

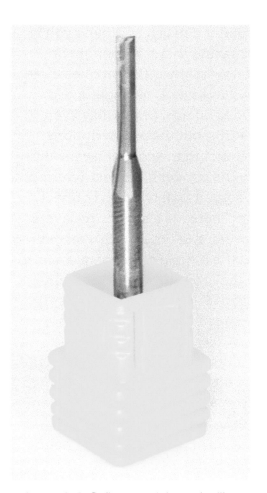

FIGURE 9-3: 2-flute straight end mill

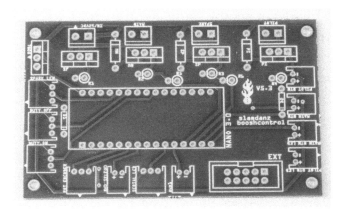

FIGURE 9-4: Boosh controller PCB

PCBs are defined in layers. Unless you're involved in making them, you might be surprised at how many different ones there are. You may have heard of single-sided or double-sided boards, or possibly even multi-layer PCBs, but these descriptions are only referencing the *copper* layers, the layers where the copper channels, referred to as *traces,* serve as flat wires to connect components. There are also ground layers, silkscreen layers, and others. Without getting into some of the more exotic layers, an intro PCB design tool like the wonderful open-source program Fritzing offers Silkscreen, Part Image, Rulers, and Ratsnest layers during design. (See Figure 9-5.)

When sending the PCB off to the fabrication house, a different set of files is required to define the operations and layers of manufacture. The Extended Gerber format (RS-274X) is a common standard for defining PCBs. (See Figure 9-6.)

For our needs, the key images are the copper traces and pads and the silkscreen. The traces are the equivalent of wires, and the pads are the copper areas where components are soldered to the board. The silkscreen is a layer that is literally silkscreened onto the board to help identify components, connections, and provide other human-readable information.

The boosh controller is a two-layer board, with traces on the top and on the bottom. For the purpose of creating an image to place on the tablet sleeve, I'm combining these. They are usually designated with different colors in a design program. (See Figure 9-7.)

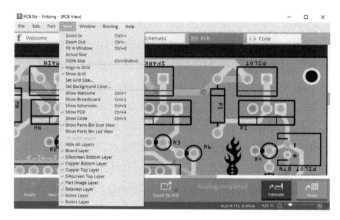

FIGURE 9-5: Fritzing layer options

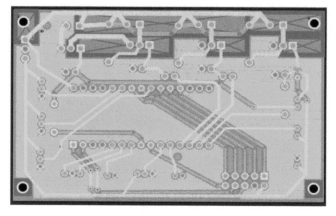

FIGURE 9-6: Gerber files for PCB production

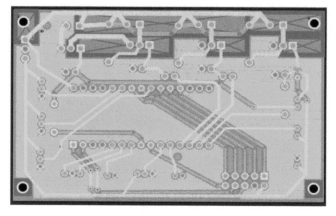

FIGURE 9-7: Copper layers with ground planes

This is a little crowded and busy, especially since there are ground planes included. *Ground planes* are large areas of copper connected to the circuit ground, and help shield the circuit and hopefully reduce electrical noise. For aesthetic purposes, I'll remove the ground planes to highlight the traces. (See Figure 9-8.)

Since we're working toward an artistic, rather than functional, target, it doesn't matter if we make adjustments to the board that wouldn't work in the real world.

The other layer of interest is the silkscreen layer. For the boosh controller, only the top layer is silkscreened. (See Figure 9-9.)

The intent is to place these on the top of the sleeve. The interesting question is how to do so. The copper and silkscreen need to have very different looks. To achieve this, the copper will be milled into the leather, and the silkscreen will be laser-burned.

Using a mill to engrave patterns in leather can be very effective. It works best when there is a method to enhance the contrast. Two ways to do this are either to dye the leather a dark color and mill away the leather to reveal a lighter lower layer, or to coat the leather with a finish, mill it away in spots, and then use dye or an antiquing gel to darken the cuts. (See Figure 9-10.)

Since we'll get a dark image when we laser-burn the silkscreen, I'll go for the first approach of dyeing, then cutting for the traces. I'll use a warm brown dye that will still

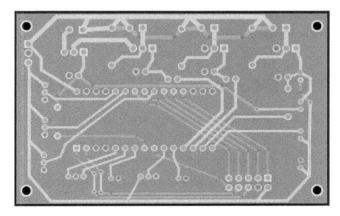

FIGURE 9-8: Copper layers without ground planes

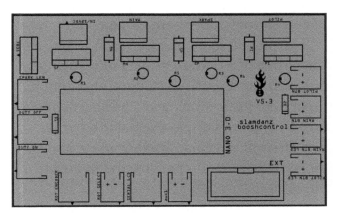

FIGURE 9-9: Silkscreen layer

FIGURE 9-10: Milled leather, light and dark

be light enough to show the laser burns, and mill it away with a 1/32" bit.

TAKING MEASUREMENTS

The tablet is basically square, but since we want to make a reasonably accurate 3D model of it, we need to determine the curves (fillets) at the corners and along the transition from front to back. The front is easy; the dimensions are published online and they verify when checked out. Determining the fillet is slightly more complicated.

It is difficult to be precise when measuring the radius of a corner. While radius gauges are available, they aren't a common item in most hobbyists' toolboxes. The standard way is to hold a right angle against the corner and attempt to visually determine where an edge stops being parallel and begins to curve. (See Figure 9-11.)

Measuring the fillet on the curve of the back is even more difficult, since it is not a symmetrical curve. The curve begins approximately 4 mm up from the edge, and then ends about 10 mm into the back. (See Figure 9-12.)

I took the start and end measurements and fitted an arc to them in Fusion 360. I'm confident that it's not perfect, but it's close enough that the model constructed from it will work as a stand-in for the tablet itself.

Since we're leaving the top open, I chose to ignore the buttons on the device. They protrude approximately 1 mm. If they're at the open top, they don't matter, and if they're inside the case, there is likely enough slack in the dimensions for them not to be a problem.

FIGURE 9-11: Measuring the fillet

FIGURE 9-12: Measuring the curve on the back

DIGITAL FABRICATION

In some ways this project should be the simplest of projects in this book. In the end, it's just two pieces of leather stitched together. If we were doing it manually, it would be as simple as overlaying the two pieces, punching the holes, and stitching. In my opinion, the effort is somewhat more complex and interesting because of the need to maintain registration for the various digitally controlled operations we'll be undertaking.

Registration is the term used to describe aligning different parts. Finding ways to register objects so that a hole punched in one will line up with a hole in another is a lot of what this book is about. We'll use some interesting approaches in this project to solve registration problems, which I hope will be useful for you in many of your own projects.

SLEEVE PARTS

Creating the bottom part of the sleeve is simple. Create a shape of the outline dimensions of the tablet, and apply the fillet to the corners. Then, offset the outline 6 mm to allow for stitching. This becomes the outline of the bottom. Create a construction line 3 mm inside the outer edge as a guide for the stitching holes by creating another offset line from the outline.

I like to add points at midpoints and where the curves begin. These help me line everything up. I added 2 mm circles at the beginning of each curve and used these to do Rectangular patterns of circles down the sides, and Circular patterns of circles along the curves. (See Figure 9-13.)

This will serve as the DXF that we will mill into the top and bottom at the same time. It will cut both pieces to the proper outline, and make sure the holes are aligned for stitching. It can be extruded to form the model of the bottom piece.

It isn't critical that you model the top piece, but I wanted to be sure I understood how much flat space I had on the top to mill. I created an extended outline of the tablet and extruded it. (See Figure 9-14.)

Assuming the curve of the case was consistent on all sides, I used the resulting body, oriented 90 degrees at the back edge, as well.

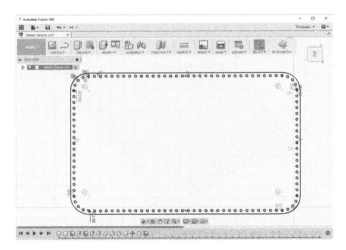

FIGURE 9-13: Bottom outline with stitching holes

DIGITAL FABRICATION **161**

I had to cut it in half since I only wanted one edge. (See Figure 9-15.)

To connect the corners, I used the Revolve tool to connect one face to the one on the other side. (See Figure 9-16.)

Once that was completed, I joined all the parts into a single body, forming the top piece. This left me with a flat piece for the top measuring 95 mm by just over 153 mm. (See Figure 9-17.)

Since we have to mold the top piece, and we want stitching holes, I decided to put those into the top piece prior to performing the molding. This allowed me to accomplish a couple of things. First, I got to make the holes while the piece was flat, and second, it provided a way for me to avoid stretching the edge during molding.

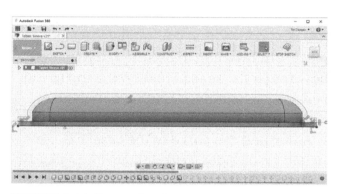

FIGURE 9-14: Top outline for extrusion

FIGURE 9-16: Revolving the corners

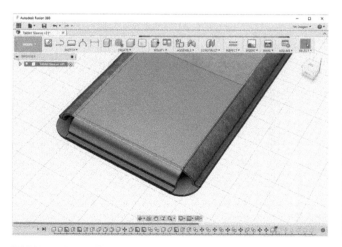

FIGURE 9-15: Back edge of top piece

FIGURE 9-17: The flat piece for the top

When molding leather, the parts where the leather is being stretched aren't the only parts affected. Other parts of the sheet of leather will expand or contract to accommodate the parts we're abusing. Leather is a big net of fibers, and changes made to one part can impact others. This is a problem for our project in that, as noted, the edge where the sleeve opens tends to want to bow up and expand during molding, which would make the sleeve loose. Pre-stitching holds it in place.

I measured the anticipated length of the top edge and included the flat parts where the outer stitching would occur. This allowed me to calculate the number of stitch holes I wanted and to create a sketch. (See Figure 9-18.)

IMAGES

The images require two files: one for the traces and one for the silkscreen. These can be traced and cleaned up in Inkscape from screenshots taken from the PCB tool. Once they are prepped as desired, they need to be scaled appropriately to fit in our target 95 mm × 153 mm space. Both of these are saved as DXF files. They're fairly large and have many points. You can use the Simplify feature of Inkscape to reduce points, but be careful; eventually that process will noticeably degrade the image.

After putting everything in grayscale and overlaying, I was able to get a decent idea of what they would look like together. (See Figure 9-19.)

3D PRINTED PARTS

The tablet model we created can be saved as an STL file and is ready to print. This gives us a waterproof replacement for the tablet, which is important during molding. Letting the wet leather press against the tablet and sit for hours while it's drying isn't great for it, or electronics in general.

The resolution isn't a big deal, but printing with a fairly high resolution provides better models of the curves. Enough infill, 40 percent or better, is also important to

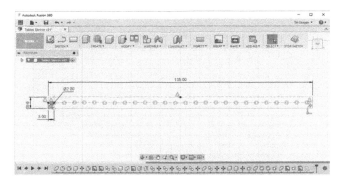

FIGURE 9-18: Reinforcement holes

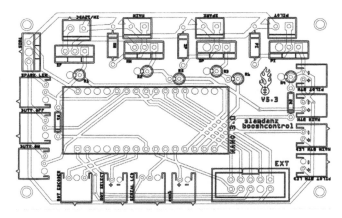

FIGURE 9-19: Top patterns

provide structural strength when we start milling on top of the model inside the leather. (See Figure 9-20.)

There is another 3D printed part that will make a big difference. To do the best possible job in molding the leather to the tablet, we need to press down all the way around the tablet as evenly as we can. To achieve this, I printed a frame that fits around the model with 2 mm to spare. (See Figure 9-21.)

This is a very simple model that just extends the outline of the tablet. I made it the same height as the tablet. It was easy to print, but may have to be placed diagonally on some printers. It could also have been cut on the CNC, but since it's pressing against wet leather, it would have to be cut from a waterproof material or coated in some way. (See Figure 9-22.)

PRE-MOLDING TOP PIECE REINFORCEMENT

Molding leather creates distortions in areas well beyond the area being stretched. One of the areas that we want to be sure we don't distort is the edge of the top piece where the tablet will slide into the sleeve. We also want to reinforce that area so that it doesn't stretch with use. We can address both issues by putting in reinforcement holes along the edge and stitching it with thread prior to molding.

To achieve this, I calculated how many holes would be needed for the top edge

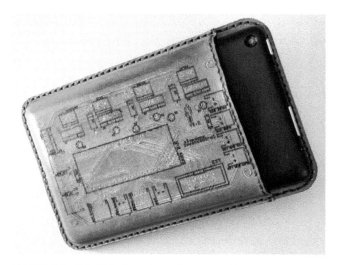

FIGURE 9-20: Printing the tablet model

FIGURE 9-21: Frame model

FIGURE 9-22: Frame and model

FIGURE 9-23: Drilling the reinforcement holes

based on its length and created a DXF file consisting of a row of 26 2 mm holes. I then made sure that one edge of the leather workpiece for the top had a straight, clean edge. I aligned this edge with the X axis of the CNC, and set X0 so that the holes would be centered on the piece. I didn't have to be especially precise, since I'd be cutting the outline in a later step after molding and just had to make sure I had sufficient width for the piece. If I was a little off-center, it wouldn't matter. (See Figure 9-23.)

MOLDING THE WORKPIECES

The top piece is molded to the shape of the tablet on three sides. This means that it must be placed correctly on the tablet model. I used masking tape on the tablet model to identify the line where the top piece should fit and marked center on it.

I cut the reinforcement holes into the edge of my top workpiece and stitched them up. (See Figure 9-24.)

Once that was stitched, I used a piece of string stuck in the stitches to designate the midpoint, which would help me align the piece during molding. I thoroughly soaked the top piece in water to prep it for molding. Soaking the 4–5 oz leather doesn't take long, and can be done under a faucet in a couple of minutes.

I laid down a paper towel on my waterproof surface to help wick away the water while the leather dried. I positioned the tablet model and placed the top piece onto it. With the top piece correctly oriented on the model tablet, I placed the frame over the top and pushed it down, using weights to maintain pressure. (See Figure 9-25.)

Molding leather is easy, but it's important to remember to never let any iron-bearing metal touch the leather when it's wet. It will

FIGURE 9-24: Stitched reinforcement holes

react with the leather and stain it a gnarly black. This might have some artistic potential, but it's just a big bummer when it ruins an important piece.

The leather will take some hours to completely dry depending on humidity. You can speed it up with a hair dryer on low, but it's better to just wait it out if you have time. Now and then, remove the frame and use a smoothing tool to remove wrinkles and tighten up the edge. You'll find that the leather can be shaped almost like frosting for quite some time. (See Figure 9-26.)

When it's fully dry, the top piece is ready for the next step.

CNC OPERATIONS

In many ways, this project isn't about making a tablet sleeve. Just as all the projects in this book are methods to explore techniques, this project is really about workholding. *Workholding* is the broad term for finding ways to keep something from moving while you abuse it with drills, saws, grinding wheels, router bits, or other stock removal tools. It includes clamps, jigs, vises, and other tools. But we're going to use a technique I've been using a lot in my shop, a homebrew double-sided tape process.

Double-sided tape is nothing new, but it's expensive, sticks poorly, and is difficult to use across the complete surface of a large item. Instead, I adhere contact paper (aka, adhesive shelf paper) to the work surface and bottom of the workpiece. I then use cyanoacrylate

FIGURE 9-25: Molding frame in place

FIGURE 9-26: Smoothing out the molding

(Super Glue) or Tacky Spray on the work surface paper and stick the two together. This is a variation on a process that was given to the world by Ben Crowe, a master luthier at *https://crimsonguitars.com/*. He uses masking tape, but the principle is the same.

The biggest challenge is that I want to accomplish four things without moving the workpiece.

1. Drill holes in the top and bottom that will match perfectly.

2. Cut the outlines of the top and bottom to match perfectly.
3. Engrave the traces of the PCB on the correct area of the top piece.
4. Burn the silkscreen pattern over the engraving and have it be correctly registered.

Trying to independently position the top and bottom and get all the holes and outline lined up correctly is frustratingly difficult. On top of that, since I've already molded the top part, it is very difficult to then engrave on top of it. So, as I often do in times of stress, I decided to make a sandwich. (See Figure 9-27.)

Since I already had a model of the tablet, I could use it as a support structure. By using multiple levels of the double-sided tape technique, I created a stack of items that allowed me to register both the top and bottom pieces for drilling and outline cutting, as well as the engraving and laser burning. (See Figure 9-28.)

I needed to place all of this so that it was true to the X and Y axes of my CNC. I wanted a spoil board, since I intended to cut slightly below the surface of the bottom piece to make sure both it and the top piece were fully cut through. Since I had a spoil board covering up the lines of my worktable, I needed some way of orienting the workpiece. To get around this problem, I taped sewing thread across the spoil board to replicate two of the lines from the worktable. (See Figure 9-29.)

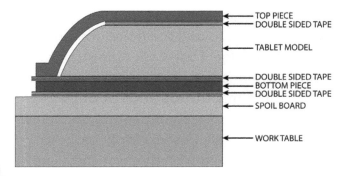

FIGURE 9-28: The first layer of contact paper

FIGURE 9-27: Workholding sandwich

FIGURE 9-29: Orienting the workpiece with threads

DIGITAL FABRICATION

(Oops!) Dyeing the Leather

The top piece needs to be dyed prior to engraving on it, so it should be done before attaching it to the CNC worktable. I realized I'd forgotten to do the dye work just before I started cutting, so I masked the surrounding area and dyed it on the table. You can dye the other piece at this time as well if you like; if not, be sure to dye it later. I also chose to dye the fiber side of the bottom piece, too, since the upper part of it will be showing. (See Figure 9-30.)

FIGURE 9-30: Dyeing the top piece

Milling and Drilling

I used two toolpaths combined into a single G-code file to drill the holes and mill the outside edge. I used a 2-flute straight 2 mm bit, since I needed a cutting length sufficient to mill the top and bottom layers combined. Instead of treating the holes like circles to mill, I used a "drill" pattern that simply used the bit that was installed to do a straight up and down action to create holes. The perfect hole size was 2 mm, so I was able to combine the drilling and outline milling into a single file to run. (See Figure 9-31.)

As extra insurance, since it was outside the fillet of the corner, I positioned the CNC at X0Y0, and lowered the Z axis until the tool dented the leather at the X0Y0 point. This served as an extra method to verify that I hadn't moved the router out of the expected position when changing tools.

FIGURE 9-31: Milling and drilling the top and bottom pieces

Once that was completed, and after re-zeroing the Z axis to the surface of the top piece, I was able to use the same X and Y registration to then run the milling pass. I switched to a ⅛" 60° engraving bit for the PCB traces.

Engraving and Burning

There are many choices when engraving on leather. The pattern I traced from the PCB provided two major options: use a routing tool to follow the lines of the traced objects, or rout out the area inside the traced objects. The ability to rout out inside the traced objects depended on the cutting width of the bit chosen. If the bit was wider than the trace I wanted milled out, it wouldn't work. I needed a bit narrower than the smallest area I wanted to mill. (See Figure 9-32 and 9-33.)

Ultimately, I chose an engraving bit narrow enough to mill all areas, and I made the choice based on time. Cutting out all the filled areas of the traces would take over four hours. Cutting just along the outlines of the traces would take 27 minutes. If I'd thought it would look dramatically better, I might have gone for the four-hour cut. But having had a PC auto update and reboot during a five-hour 3D printing job recently, I chose to take the safe way out and go with the 27-minute approach. (See Figure 9-34.)

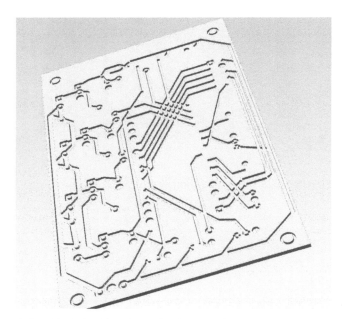

FIGURE 9-32: Preview of milling entire traces

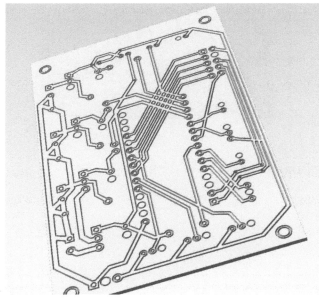

FIGURE 9-33: Preview of milling the outline of traces

With that completed, a slight offset allowed me to then use the laser attached to my router to burn the silkscreen pattern onto both the untouched and routed leather. This is the way PCBs work: the silkscreen layer is applied on top of the traces. (See Figure 9-35.)

With all the CNC work completed, the sandwich can be peeled apart and the leather can go on to the next stage of leatherworking finish.

FIGURE 9-34: Milling the PCB traces

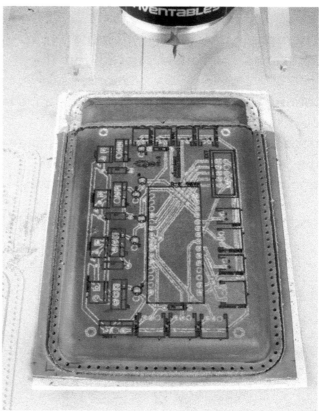

FIGURE 9-35: Burning the silkscreen layer

LEATHERWORKING

The traditional leatherworking for this project is very simple. We've already dyed the work (if you haven't, complete the dyeing now). What remains is to assemble the project.

CLEAN UP AND TRIM

Since we had the pieces right-side up during the milling operations, the edge of the skin side of the leather isn't as pretty as in previous projects. This can be cleaned up by careful sanding, a judicious use of embroidery scissors, and, in some cases, careful use of a utility knife to cut the edge back clean.

You may also find that there are little tags sticking up from the engraving. This is due to the tightness of some of the traces, which left very narrow strips of pieces of surface leather. These dangling bits can be snipped off with embroidery scissors. We'll use a smoothing tool to push down everything else and add a heavy layer of finish coat to lock everything in place.

FINISH COAT

After applying the first finish coat, I used a smoothing tool to press down any of the little tags and bits of leather. The finish I'm using is an acrylic-based product, and, like acrylic paints, it serves as a decent glue. After smoothing everything down into the still-damp finish, I let it dry, and repeated the process with four more finish coats.

I also applied a black edge coat.

SEWING THE TABLET SLEEVE AND FINAL STEPS

Stitching the two pieces together is a very simple task, especially if you leave the stitches from the top piece reinforcement holes in place and treat that as a separate section. With this approach, I started at the bottom center and stitched all the way around the project with three overstitches at the end. For additional reinforcement, I added a stitch around the outside edge at the last hole of the top piece on each side. (See Figure 9-36.)

FIGURE 9-36: Stitching the project

You may decide that the edges are rougher than you like. If this is the case, you can take a skiving blade and lightly run it along the edge to clean it up. I also recommend a nice edge coat on a project like this. Edge coats can feel risky if you try doing it with a brush. I highly recommend using an edge roller to get great-looking edges without getting the paint on the main surfaces. (See Figure 9-37.)

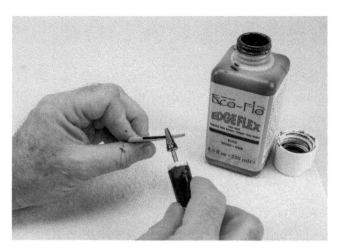

FIGURE 9-37: Edge coating with an edge roller

THE RESULTS

The PCB traces are subtler than many folks might want them. If that's the case for you, I recommend laser engraving both layers, rather than doing an overlay pattern like I used. This project is really a set of techniques to show the possibilities of using your 3D printer in conjunction with the CNC. I'm sure you'll come up with exciting new ideas for the motif that fits your project the best. See Figure 9-38.)

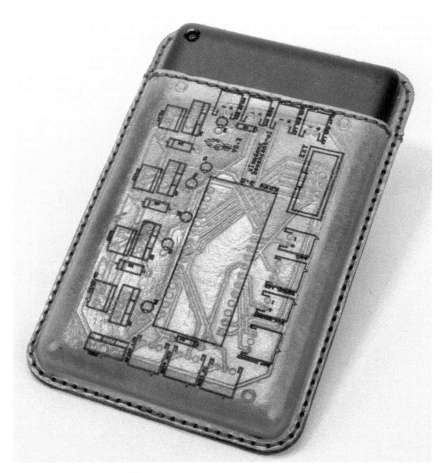

FIGURE 9-38: PCB tablet sleeve

ELDER GODS BELT POUCH

Howard Phillips (H. P.) Lovecraft welcomed the 20th century with a vision of dark eternity. From 1917 to 1936 he wrote a range of stories that included the Cthulhu mythos and stories of the Elder Gods. His tales of unspeakable horrors finding their way into the world around us have become a cultural mainstay in books, comics, movies, and games.

DESIGN

In tribute, we will summon Cthulhu from his watery R'lyeh to adorn our next project, a belt pouch. But not just any belt pouch: one specifically designed to accommodate a special item that has taken much of the making world by storm, the mighty Altoids® tin.

Altoids tins have become ubiquitous as containers for everything from electronics projects to survival kits to the fuselage of drones. Many responsible smokers will carry one to police their butts. I have multiple tins I use as sewing kits, tool kits, and even to hold my curiously strong mints.

These containers are so popular that I decided that this case should hold not one, but two of these useful tins. Luckily, many sites on the Internet, such as *GrabCAD. com*, have prepared models that can be downloaded as a starting point for modeling the case.

TAKING MEASUREMENTS

I made a copy of the model that I imported into Fusion 360 and stacked them side by side. To make things easier, I then created a basic cube around them that represented the space I wanted inside the case. This cube was 96 mm × 45 mm × 62 mm. (See Figure 10-1.)

The pattern could be adjusted to accommodate only a single tin, or to allow you to stack more than two tins in a vertical orientation. The approach we're using is straightforward for any cubic internal dimension.

THE BASIC LEATHER PATTERN

This is a belt pouch, so it will have a belt loop; but, with that exception, I wanted to cut this as one piece of leather. That's not always the most desirable approach, because often the available leather or the constraints of the pattern will dictate multiple pieces as the solution. But I want to use this project to share how to take a 3D design and unfold it into a flat 2D pattern that we can cut as one piece.

One of the areas that is very active in CAD work, but not well exposed to the DIY

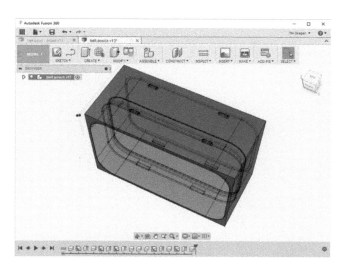

FIGURE 10-1: Stacked tins

176 ELDER GODS BELT POUCH

and Maker communities, is sheet metal CAD. Commercial tools with high price tags rule this important space. The idea is that you can design a product to be formed from sheet metal, and the software will unfold it so that you can stamp out a single piece. There are a few less-expensive CAD packages that have some features or plugins that do this to varying degrees, and software for working with paper models does many similar things.

The activity of unfolding sheet metal designs has been around for well over a century. I love to collect old books from technical schools of the 1920s and 1930s, and many include beautiful drawings with supporting math to undertake this operation. (See Figure 10-2.)

We're going to use a simple hybrid method that you can do in pretty much any 3D CAD program. It's not as precise as the high-end sheet metal CAD, or as complex as the manual method, but it's easy to do and the results work well with leather.

The case we want is simple. It has leather on all six sides of the internal cube and a flap over the front. We'll need to stitch the pouch together, so the sides at the back and bottom are slightly extended and overlap. We'll use snaps on the front for closure. The snaps and the stitching both require holes and we'll use the CNC to create them. (See Figure 10-3.)

To help visualize the steps, I'm going to jump ahead and show you the resulting pattern we want to end up with. (See Figure 10-4.)

The interesting question is how to get from one to the other. You can more or less accomplish this by measuring the size of each face and then organizing them as a flat continuous surface. The risk is that the stitching holes won't line up correctly. This

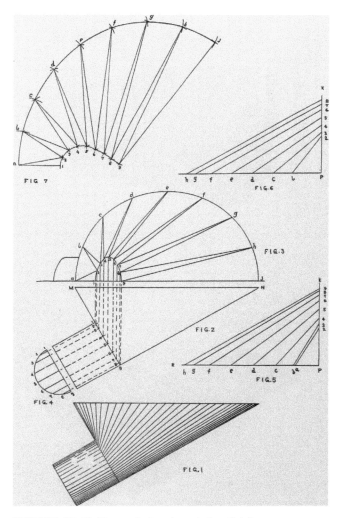

FIGURE 10-2: Manual sheet metal unfolding IMAGE COURTESY OF WIKIMEDIA COMMONS

isn't a problem when doing the pattern-making by hand; you just wouldn't punch the stitching holes until you had the pattern all folded up. But we want to use the CNC to do most of the work, so we need to be a bit more precise.

I started with the basic model and made sure I had my stitching and snap holes already in place. It doesn't really matter which way you put the model together, but you do need a way to find the proper rotation point. I did this by splitting the bodies of my model using angled planes, effectively creating mitered joints between the pieces. (See Figure 10-5.)

This is easy to do; just select the face you want to split and create a sketch. Draw a line where you want the cut to occur, then select this line and extrude it so that it passes through the body. (See Figure 10-6.)

With the Split Body function, you can use the miter plane like a knife and cut the body with it. (See Figure 10-7.)

Doing this allows you to choose three location points to rotate (unfold) from. You can do it at the top if you're intending to groove the inner surface. You can do it in the middle if you're just making a bend that will stretch the outer surface and compress

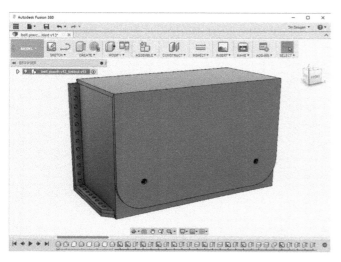

FIGURE 10-3: Target design

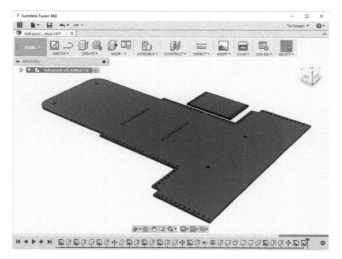

FIGURE 10-4: Target pattern

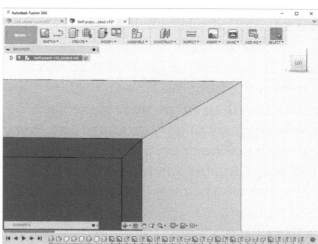

FIGURE 10-5: Splitting the model with a miter

the inner surface. You can do it at the bottom if you're molding a bend that will primarily stretch the outside of the leather. (See Figure 10-8.)

Using these processes, you can see the sequence of unfolding I created for my model in the diagram shown in Figure 10-9.

After rotating each section, I combined it with the other section in the same plane. This served to reduce the parts count at each fold. Steps 6–8 are the hardest, because they are the stitching edges and need to be correct. You'll notice that I didn't unfold both sides. The process was complex enough that I did it correctly on one side and then, in

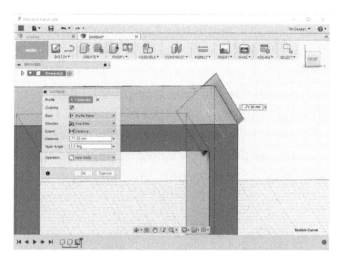

FIGURE 10-6: Sketching the miter line

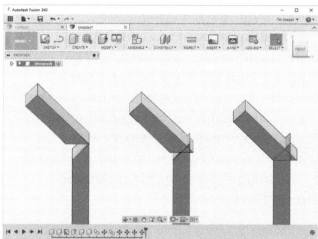

FIGURE 10-8: Different rotation points

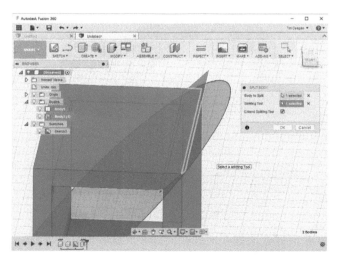

FIGURE 10-7: Splitting the body along the miter line

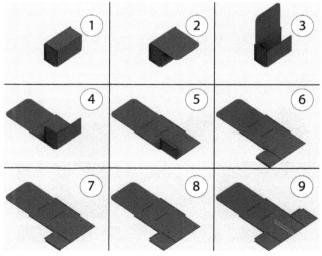

FIGURE 10-9: Unfolding sequence

DESIGN **179**

Step 9, created a midline plane that allowed me to "mirror" the completed side.

CREATING CTHULHU

I couldn't find artwork for Cthulhu that I liked and was available in the public domain. So I created my own. The process is useful for making your own designs, so I'll briefly outline what I did.

Cthulhu has, more or less, an elephant's skull, a cuttlefish's tentacles, and bat wings. Starting with a Google Image search, I used the Search Tools to find images that were "Labeled for reuse," meaning that there is no copyright to worry about. In truth, it's always worth following through to the image to verify this status. (See Figure 10-10.)

I ended up with two images that had the pieces and parts I wanted. (See Figure 10-11.)

I roughly cut the bat wings off and rounded the squid head. At this stage, I only needed a rough image—it doesn't have to look good yet. (See Figure 10-12.)

With my basic image composed, I then traced the lines of interest. This is done using the Pen tool in Inkscape or Illustrator. First, create straight lines (you can convert them to curves later). (See Figure 10-13.)

There are excellent tutorials about this process on the Internet that go into detail on how to use the Pen tool effectively. Once you have the areas outlined, then convert the anchor points to curves (unless the spot should be a hard corner). You can taper lines

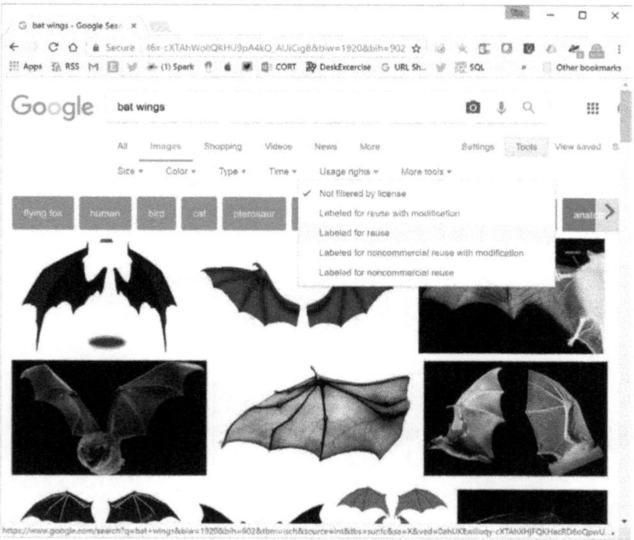

FIGURE 10-10: Google image search license tool

FIGURE 10-11: Squid and bat

to get a more artistic effect. Thicker lines look better on the outside. (See Figure 10-14.)

I wanted a little more punch, so I added the Sigil of the Gateway associated with Cthulhu. I first added it on the forehead, and then again in the background to create a frame for the image. (See Figure 10-15.)

Learning the mechanics of this process can take practice, but I hope your takeaway is that you can make your own images to suit your needs from parts and pieces. Seeing your own art on your projects adds encouragement that using someone else's art never will.

FIGURE 10-12: Cthulhu parts

FIGURE 10-14: Tracing complete

FIGURE 10-13: Rough tracing

FIGURE 10-15: Final image

DESIGN

DIGITAL FABRICATION

We have a single CNC cutting/drilling pass and a laser-burning pass in this project. Both pieces of leather, the pouch and the belt loop, are cut and drilled from the same workpiece. With this project, we'll also do a tool change and cut the folding grooves using a V bit.

PREPARING THE FILES

I hope you're starting to feel comfortable with the workflow of producing these projects. The types of files we're producing are very similar for all of them. While I'm sharing the toolset and file types that I'm using, what's important is understanding the actions to undertake rather than the specifics of file type or software.

Leather

Once we have the model unfolded in our CAD program, we'll need to lay the belt loop piece in the same plane as the other piece. We can use the face of any object in Fusion 360 to create a sketch. The sketch is a 2D drawing of that flat face, and can be exported as such. This is how we convert from a 3D image to the 2D DXF file format. Additionally, I've added lines to represent where we'll want the folding grooves to be cut. (See Figure 10-17.)

From the DXF we need to create four different toolpaths. The first is a drilling toolpath. I'm using the same 2 mm straight flute bit used in the last project. This means that

---

Laser Alternatives

If a laser isn't available, you can still use the process we undertook with the tablet sleeve to engrave directly into the leather with a cutting bit. If you go this route, save that step until after you've done the dye work.

If you want the engraved area to be dark instead of light, dye it, and add a couple of finish coats to the leather. Then, after you've finished the engraving pass you can use an antiquing gel. This is a dark gel paint that you smear across the entire part. The leather protected by the finish coat won't soak up the gel paint, but the engraved area will. Some amazing and incredibly fine effects can be created this way. (See Figure 10-16.)

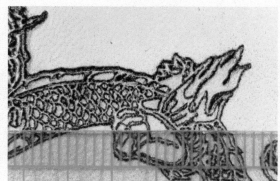

FIGURE 10-16: Section of engraved and antiqued image on leather (scale in mm)

I can use a drill function to do a straight up-and-down to make a 2 mm hole.

The second toolpath is for the snap holes. These are ⅛", larger than 2 mm, so the path for the bit is an inside circle cutting out a space larger than the bit itself.

The third toolpath cuts out the pieces. This is the outline of both the pouch and the belt loop. It is an outside cut to preserve the dimensions of the parts.

All three of those toolpaths use the same bit and cut all the way through the leather. My leather is approximately 1.8 mm thick, and I set the cutting depth to 2 mm. But the last toolpath is for cutting the folding grooves. We want a V-shaped cut that only goes about ⅓ of the way through the leather. We'll select those lines and create an engraving path that follows the lines directly. The depth of cut will be set to an appropriate depth. For the leather I was using, this was 0.8 mm. (See Figure 10-18.)

Since I could have the first three paths cut sequentially without a pause, I combined them into a single G-code file. The last toolpath uses a 90° V bit, so I need to change tools and re-zero the Z axis. I put it in a separate G-code file that I would run once those steps are complete.

Cthulhu

I wanted the Cthulhu model to have blackened wings, so tracing the image wasn't going to work. Instead, I imported the image into my laser program, T2Laser, and set it up to do a black and white image. For image work, the laser operates in raster mode. The software sets the laser up to move in incremental rows, turning the laser on and off. Televisions used to work this way, scanning across the screen; most printers still work this way, creating the image one line at a time.

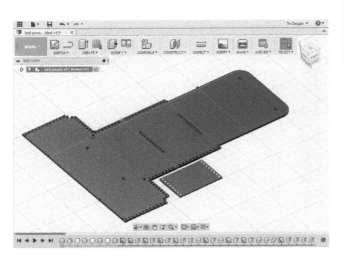

FIGURE 10-17: Model ready

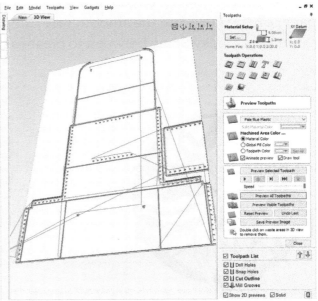

FIGURE 10-18: Preview of the four toolpaths

The program I'm using (and many others) allows the user to select horizontal or diagonal scan lines. I find that the diagonal scan lines aren't as obvious with most images. The pouch is 100 mm wide and I wanted the image to nearly fill the width, so I resized the image to 95 mm. (See Figure 10-19.)

I selected the appropriate power level and speed for my laser. I'm using the 15 W laser on my CNC, so this was 2250 mm/min and level 64 (out of 255) for power.

PREPPING THE WORKPIECE

We need at least 275 mm × 200 mm for the leather, but it helps to have a little slack. I cut out a 300 mm × 225 mm square of 4–5 oz leather to use as my workpiece. (See Figure 10-20.)

I used the same workholding method previously described: sheets of contact paper on the table and the smooth face of the leather, Tacky Spray between to create a custom-sized piece of double-sided tape. I positioned the X0Y0 point a few millimeters inside the bottom left corner. This allowed a safety margin for any deviation from the XY registration—the workpiece can be slightly off-kilter and I'll just use the extra width and length to accommodate.

CNC WORK

With the 2 mm 2-flute straight grooving end mill in the chuck and the Z axis zeroed, the three operations combined into the first G-code file can be executed. (See Figure 10-21.)

When the X0Y0 area is going to end up outside the project's final cut area, I like to carefully lower the router and bit so that it makes a small indentation in the leather at X0Y0. This is useful in case something happens that makes you lose the registration of the head. Trying to get the bit to start at the same place without a registration mark is

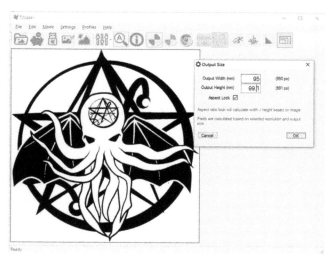

FIGURE 10-19: Generating the image toolpath

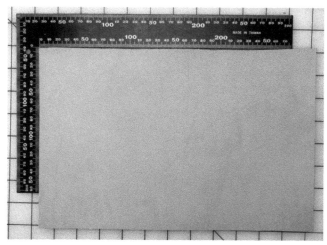

FIGURE 10-20: The leather workpiece

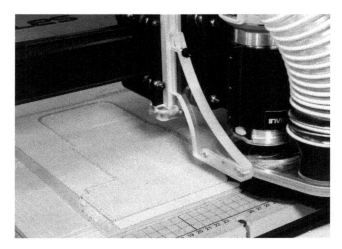

FIGURE 10-21: Holes and outline completed

frustrating. With a small dent in the leather, you can get the head back to a place very close to the original X0Y0.

When you're ready for the second G-code file, move the Z axis to a comfortable height and change the tool to a 90° V bit. Be careful not to move the router when changing tools, because this will make you lose registration. Unless you made the mark described in the previous paragraph, you'll have to extrapolate from some other hole or cut mark to get back to the X0Y0 you started with.

Zeroing the Z axis

Knowing where the Z0 position is may seem straightforward, but it changes every time you change bits, move the router or spindle in its mount, add a spoil board, or change the worktable. Some operations are defined in negative terms, with Z0 at the top surface of the workpiece rather than the table surface. Re-zeroing the Z axis is something you will need to do on a regular basis.

It is common with most CNC machines to have an electronic Z-probe that makes finding Z0 easy. These units are very simple. Essentially, they turn the bit and table into a switch. When the bit just barely touches the table, the switch closes and the machine declares the position Z0.

The bit is typically electrically conductive, but many worktables, especially those with spoil boards, are not. To get around this, most Z-probes use a finely machined piece of metal set on the work surface as the other contact of the switch. This piece of metal, often referred to as a *puck*, must be flat on two parallel sides and of a known height. (See Figure 10-22.)

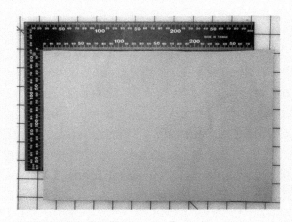

FIGURE 10-22: Electronic Z-probe

DIGITAL FABRICATION **185**

> A wire is connected to the bit and another to the puck. G-code commands exist that will execute a zeroing operation. The command set I use is:
>
> `G90;G21;G38.2 Z-75.00 F125.00;G92 Z15.00;G0 Z20.00`
>
> G90 sets the system to absolute positioning.
>
> G21 sets units to millimeters.
>
> G38.2 Z-75.00 F125.00 tells the machine "probe toward workpiece, stop on contact, signal error if failure." The Z axis is the designated axis in this command and it will probe no more than 75 mm down. The feed rate is set with F125.00, which is 125 mm/min.
>
> G92 Z15.00 is a *coordinate system offset*. Since my puck is 15 mm thick, the probe will stop 15 mm above the surface I'm zeroing on. This command takes the height of the puck into account so that Z0 is set to the surface.
>
> G0 Z20.00 then moves the Z axis to Z20 as my parking location when the operation is complete.
>
> Makers managed to zero the Z axis on machine tools long before electronic probes existed. I have a wonderful Sherline vertical mill that I zero manually. With no electronics, the trick I was taught was to use ultra-thin cigarette rolling papers. A rolling paper is held lightly to the work surface while the bit is lowered. As the bit comes close, start sliding the paper back and forth. When the paper is grabbed by the bit, you are at zero. This process is also used on older 3D printers that don't have an inductive sensor.

The second G-code file will run the V bit across the lines defined. Between small variations in the work table, spoil board, and leather, it is highly likely that the grooves won't be at a consistent depth. But they should be close enough that they will serve as a guide for a quick pass with a hand groover to improve them. (See Figure 10-23.)

FIGURE 10-23: Grooves cut

LASER WORK

Since we cut from the backside in the previous operation, we'll need to remove the leather to position it for laser engraving on the front. We don't need to adhere the leather to the table for laser work, but we do need to orient it. I set up the art so that X0Y0 would be at the lower left corner of the flap. There isn't any leather there because we used a radius on the corner. But it's an easy spot to find on our table.

The reason that this is easy is because we cut into the spoil board on our initial CNC operations. This means there are lines milled into the spoil board surface that we can use both to align the workpiece and to determine X0Y0.

With the laser on a low power setting (but you still *must* use your safety glasses), position it so that it is at the X0Y0 mark created by extending the left side and bottom edges of the flap. Position the workpiece directly over the cutout area still on the spoil board and execute the file. I chose to burn this on dry leather, but you can lightly mist it with water if you want. (See Figure 10-24.)

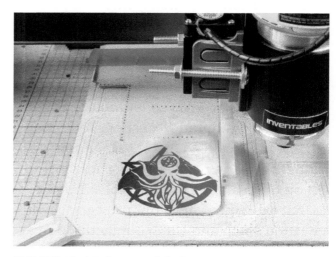

FIGURE 10-24: Burning Cthulhu into the leather

LEATHERWORKING

When the laser work is complete, don't be alarmed if the image appears overexposed. Unless you overpowered the laser, the excess black is the charred carbon from the engraving. Before you move onto the next set of operations, take the leather and hold it under running water. It will help to lightly rub it with a very soft toothbrush or other soft brush to get the carbon off. You should discover a much more nuanced image when you're done. If not, you'll have to go back to the drawing board and run tests with your laser until you understand the right power level.

CLEAN UP AND TRIM

I like to give the edges a light sanding with an emery board to clean up some of the dangling fibers. You may find that it's just as easy to place the piece on a cutting mat and run a utility knife carefully along the edge to cut away stray fibers. (See Figure 10-25.)

Advanced leatherworkers often spend a great deal of effort on edges, beveling or skiving them as desired. I tend to like a clean, hard edge, so I just want mine smooth.

As noted earlier, you may find that you want to improve the depth of some of the grooves. If so, go over them with your hand groover.

With the grooves cut and leather wet, I wanted to set the folds by letting the pouch dry in the configuration desired. To do this, I did two things. The first is to take some leather stitching needles with a little thread left on them and tie the end holes together so that everything lines up. You can also do a basting stitch and stitch lightly through a few holes if you like.

The second thing is to clip the edges together. If you must use binder clips, make sure you provide them with leather jaws. Metal and wet leather are a bad mix, and the leather will be badly stained.

I've come to prefer small plastic clips used for quilting and sewing. They're easy to position and won't stain the leather. (See Figure 10-26.)

COLOR AND FINISH

Dyeing the leather is very simple; don't forget to include the belt loop we also cut earlier. Choose your favorite dye color and then add a couple of coats of your favorite finish. I almost always end up with Canyon Tan and Super Shene, but new options are always arriving. (See Figure 10-27.)

Once the finish is dry, I used black edge paint on all edges. Be sure to remember the belt loop; it's frustrating to sew it on and then remember that you wanted it to have a black edge, as well. (See Figure 10-28.)

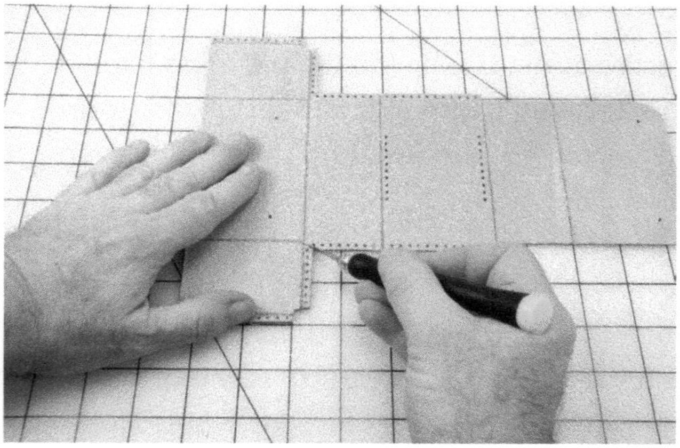

FIGURE 10-25: Cutting away fibers

FIGURE 10-26: Pouch tied and clipped to mold

FIGURE 10-27: Dyeing the pouch

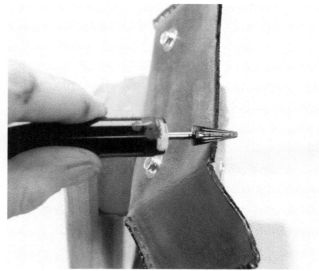

FIGURE 10-28: Edge painting

SETTING THE SNAPS

I used brass-colored line 20 snaps. Studs would have worked, as would most any closure. I didn't want the edges to curl up, so I used two snaps for a secure close. (See Figure 10-29.)

STITCHING THE POUCH

Stitch the belt loop onto the back of the pouch first. I added an extra hole on either side of each stitch row for a reinforcing stitch. This will help keep the belt loop from peeling away from the pouch with use.

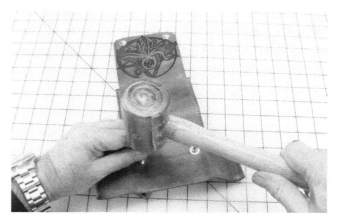

FIGURE 10-29: Setting the snaps

Stitch the overlapping sides together on the left and right of the pouch and you're done!

THE RESULTS

Whether you're eager to display your enthusiasm for the Elder Gods, or just want to add storage to your utility belt, this project is a great one to undertake. You can change the design to hold a single tin or go for a double-wide version, but no matter how your project turns out, you'll have a distinctive and useful way to store almost anything in style! (See Figures 10-30 through 10-33.)

FIGURE 10-31: Back

FIGURE 10-30: Front

FIGURE 10-32: Open

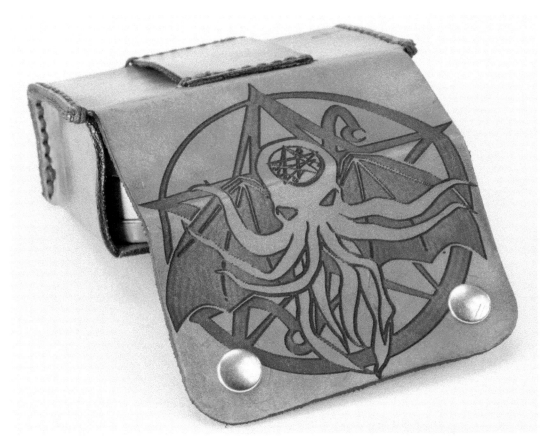
FIGURE 10-33: Top

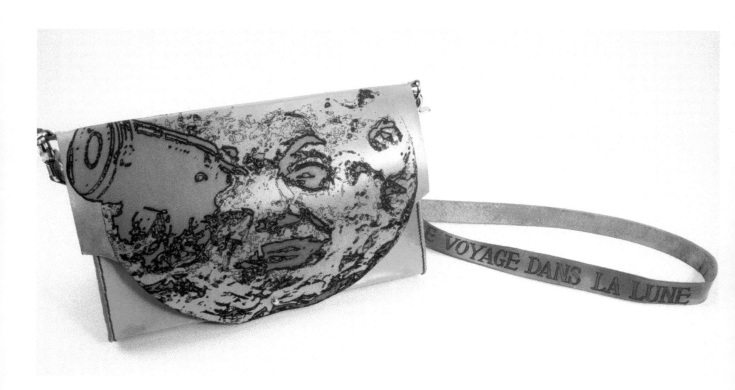

LE VOYAGE DANS LA LUNE SHOULDER BAG

Leather has a surprising ability to evoke nostalgia. Perhaps because of the many senses it engages, or perhaps because we associate it with old and heirloom items. When the motif is also evocative of nostalgic themes, the resulting leatherwork is often powerfully attractive. As a science fiction fan, the opportunities for leatherwork abound, but for this project I wanted to pay homage to the earliest science fiction film, Georges Méliès's *Le Voyage dans la Lune* (translation, *A Trip to the Moon*). This 1902 silent film astonished viewers with special effects never seen before. Comical by modern standards, Méliès's masterwork set the stage for everything that has come since.

DESIGN

The iconic image from the movie is the face of the moon with the voyagers' ship stuck in one eye. This will be the dominant image for the project. (See Figure 11-1.)

Other images from the film stick out, as well. There is a scene with moon maidens that I particularly love. (See Figure 11-2.)

The voyagers' ship also stands out and provides an interesting shape for the bag itself. (See Figure 11-3.)

The images from the film are frustratingly low-res due to the poor quality of the copies and the quality of early movie film stock. Nevertheless, the images are viable for extraction. For this project, I'll use a variable power laser mode to engrave grayscale images onto the leather.

Engraving images with different shades of gray can be tricky. The laser's power can be controlled via *pulse width modulation (PWM)*, which turns the laser on and off in a repeating cycle, to provide variable levels of output. The ratio of on time to off time becomes the power level. This ratio is often referred to as the system's *duty cycle*. (See Figure 11-4.)

Theoretically, it's possible to create a wide range of grays with the laser. However, as the wise woman says, "The difference between theory and practice is that in theory there is

FIGURE 11-2: Moon maidens

FIGURE 11-1: *Le Voyage dans la Lune*

FIGURE 11-3: The ship

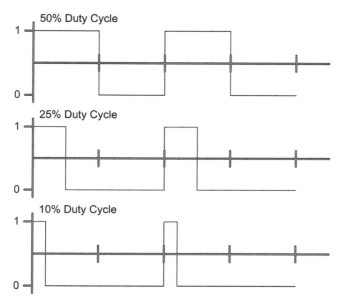

FIGURE 11-4: Pulse width modulation—duty cycle

FIGURE 11-5: Normal and posterized image

no difference." The reality is that it isn't easy to make a wide range of grays on leather by charring it with the laser. And since engraving in this manner is a raster operation (scanning rows) rather than a vector (tracing lines) one, it takes considerably longer. It's not hard for a sizable image to take hours to burn.

The easiest way to make a good set of tradeoffs is to reduce the number of gray levels desired. I find that it's hard to show more than five on the laser. Graphics programs like GIMP or Photoshop can reduce the number of levels with a couple of methods. One easy method is to posterize the image and select the desired number of colors. Since we're working with a grayscale image, this becomes the number of grays we'll try to burn. (See Figure 11-5.)

Setting the speed and power levels requires experimentation. Lasers of different powers will have different optimal settings for burning grayscale images. My 2.5 W laser is best at 200 mm/min and 180 (out of 255, about 70 percent) power. My 15 W laser is best at 5000 mm/min and 64 (out of 255, about 25 percent) power. Those settings are for dry leather, but wet leather has other characteristics. I usually prefer using the laser on wet leather, but I've found that the grayscale work ends up looking better on dry leather.

The bullet-like shape of the voyagers' ship provides an interesting opportunity. Rather than a bag with a rectangular cross section, the space ship could be the side of the bag. We can use similar techniques to the ones we used to sew the sides of the belt pouch.

However, since some of the ship's sides aren't straight lines, we'll have to use different techniques to align the stitching holes. (See Figure 11-6.)

Using the image of the moon with the ship stuck in its eye on the front flap of the bag, the desired outcome looks like Figure 11-7.

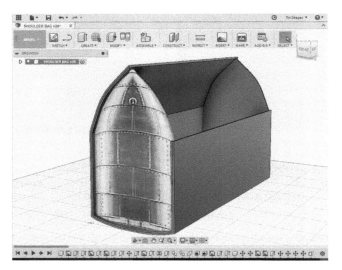

FIGURE 11-6: Ship as bag side

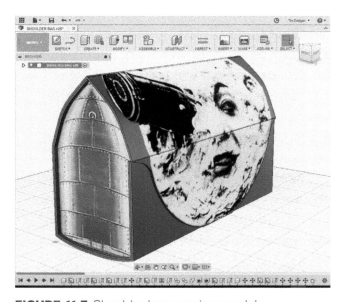

FIGURE 11-7: Shoulder bag preview model

The main body, which wraps around the sides, was a large rectangle with one edge cut into a curve to match the moon's edge. (See Figure 11-8.)

To determine the dimensions of the rectangle, I measured the length of the sides of the side panel. Curved lines can be frustrating to measure until you realize that Fusion 360 will do it for you if you simply select the line. (See Figure 11-9.)

Adding stitching holes to the main body of the bag is straightforward. After adding the fold lines, I used the Rectangular Pattern function of the Sketch tool to add all the holes (avoiding the area where the fold lines would be). Using the image I wanted to burn as a "decal," I trimmed away the front edge so that it matched the curve of the moon. (See Figure 11-10.)

Mapping the stitching holes to the bag sides was the biggest challenge of the design. I began by creating extensions to

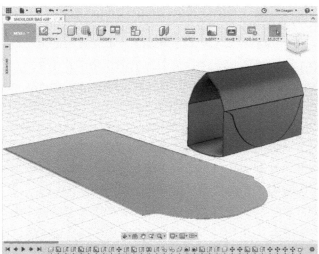

FIGURE 11-8: Main body, wrapped and flat

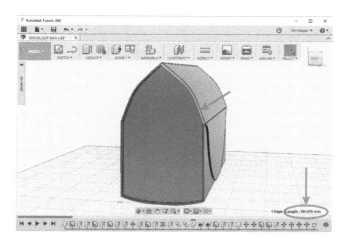

FIGURE 11-9: Measuring curved lines

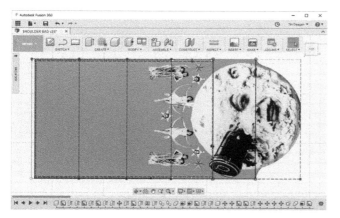

FIGURE 11-10: Main body layout

the sides at 90°. These would extend the edges 5 mm out from the body. This was easy for the flat parts that will form the front and back lower sections, but required a little more work for the curved sections at the top back section and the bottom of the bag.

The challenge is that the extensions need to be the same length as the edge. To do this, I marked the center point on the edge I wanted to extend and created a 5 mm perpendicular line at that point. This allowed me to have confidence in centering the extended line. I copied the original line and placed it so that the center point mated with the end of my perpendicular line. I then created lines at both ends to create a closed rectangle. (See Figure 11-11.)

The next challenge is placing the stitching holes. I needed them to match up with the holes I'd created for the main body. The flat sections at the sides of the lower front and back of the bag are easy; I simply transferred the set of holes created for the appropriate piece of the main body, being sure to place them centered in the extended sections.

The two curved sections were a different problem. I decided to do the bottom properly and cheat on the top. For the bottom section, I needed to take a straight line of holes from the main body and create a matched set of curved holes on the side. I created a copy of the curve in the middle of the 5 mm

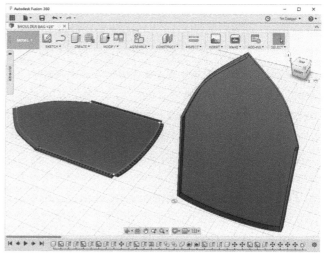

FIGURE 11-11: Extending the sides for stitching

DESIGN **197**

extension and laid out the line of bottom stitching holes directly below it. I then created lines that connected each stitching hole on the straight line to the place directly above on the curved line. I could then create stitching holes at each location. (See Figure 11-12.)

This approach isn't perfect, but it works within the tolerances required for matching the holes for stitching. The top curved section was a bit more challenging, especially since I wanted to be able to attach a strap in that area later. My cheat was that I created holes in the main body for three rivets and did nothing on the sides. Later, when assembling the bag, I'll use the holes on the main section as a guide to punch holes on the side section. This is considerably less complicated than working everything out in CAD. (See Figure 11-13.)

It's always useful to remember that all your traditional leatherworking techniques are still available to you. When using digital fabrication techniques, it can be seductive to do everything possible with computers. But sometimes it's a great deal less work to just punch a hole by hand. On this project, I've made an effort to use simple manual techniques to make things quicker and easier than doing a 100 percent digital effort.

The last design component is an inside pocket. I decided to use pigskin to create an internal folded section that would allow special items to be segregated from the main

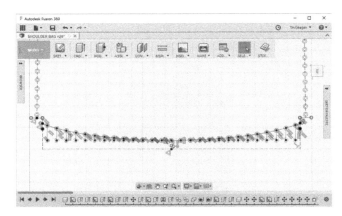

FIGURE 11-12: Mapping stitching holes onto the curved bottom section

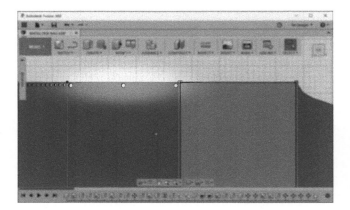

FIGURE 11-13: Rivet holes

area. Pockets like this are easy to add to any bag project; just fold a piece of material and sew it between the edges of the main leather parts.

I chose pigskin because it's thin and strong. It's usually used for lining bags, but works well as a simple internal pocket and isn't difficult to stitch in between the main body and the side. Since the edges could tear with extended use, I created holes to stitch a fold along the top of the front and back. I copied the holes from the lower back section of the main body (the location I would sew the pocket into) and simply doubled everything to create the pattern. (See Figure 11-14.)

The last component of the shoulder bag is the strap. Straps are extremely easy to create, so I decided, except for laser-burning the title of the movie, to create the strap by hand using standard leatherworking techniques.

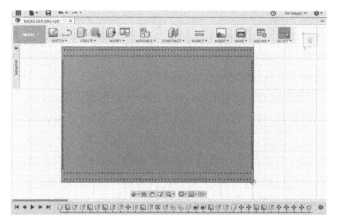

FIGURE 11-14: Internal pocket pattern

DIGITAL FABRICATION

I used to say that I collected hobbies for a hobby, but really my interest is in skills acquisition. As a result, I have to confess that I'm often more interested in the problems that a project offers than the final result of the project itself. This project pushed me to find ways to map curves in the bag's shape to flattened DXF files and to find ways to get better at shaping the leather to match the outline of images. These efforts were primarily handled in the creation of the digital files, which made the fabrication of the bag fairly simple.

CUTTING AND DRILLING

We need to produce the main body, two side pieces that are mirror images, and the pigskin pocket. The main body is the largest piece at 641 mm × 300 mm. This may be too large for many CNC work areas. If so, the three straight edges of the piece can be cut by hand. The resulting piece can then be positioned on the CNC to cut the curve, holes, and fold lines. My table was large enough for the whole piece, but I still did the straight cuts by hand. (See Figure 11-15.)

To make use of smaller pieces of leather, you could cut each of the sides individually. Each of these would need a workpiece at least 205 mm × 125 mm. I combined the sides into a single cut pattern and used a piece that was at least 270 mm × 215 mm.

Both the main body and the sides needed the outline cut, the holes drilled, and the fold lines engraved partway into the back side of the leather. All of these operations could occur with the leather held face down using the same double-sided tape mechanism from previous chapters. The outline and the holes can be combined into a single G-code file, but once again, we have to make a tool switch (and a new G-code file) for a V-cutting bit with the fold lines.

These operations are functionally identical to previous projects. For me, the only difference in this case was that I changed to a foam spoil board. This ¾" high-density foam is very level and is easy to make registration marks onto. It's too soft for use as a spoil board for heavy machining operations, but for leather it does very well.

Prior to beginning the operations, I used a drag knife (though any small bit or marker would work), to create a straight line on the spoil board on both the X and Y axis. I used the intersection as X0Y0 for all operations and used the marked lines to align the workpiece.

Alignment of the pieces is important because later, you'll need to flip the leather over for laser burning. We need a reasonable alignment of the laser pattern (a millimeter or less) for the project to look good. The outline cutting and hole drilling are set up to go through the leather. This helps by marking outlines of the workpieces.

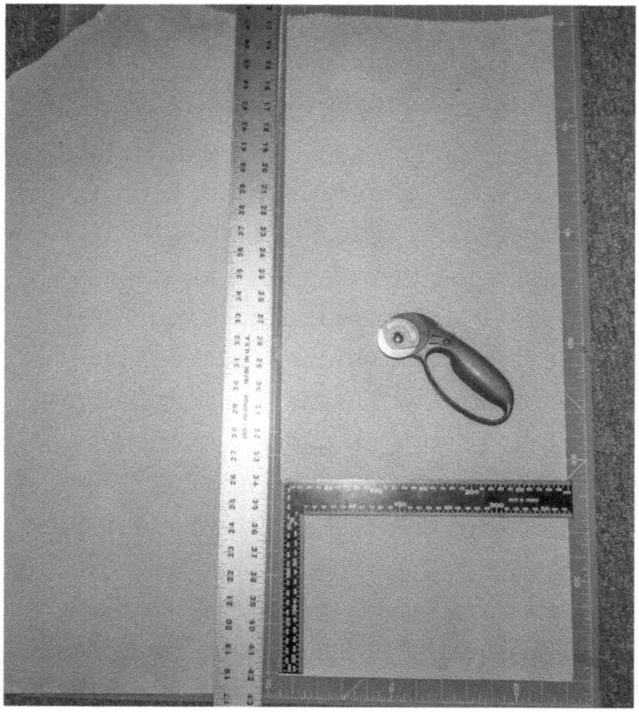

The pigskin for the pocket isn't a good candidate for cutting with anything other than the drag knife. Since the piece is a simple rectangle, I cut it out by hand and only performed a drilling operation for the required holes. (See Figure 11-16.)

BURNING OPERATIONS

With the main body flipped and aligned to the three straight sides cut into the spoil

FIGURE 11-15: Manually cutting out the rough main body

FIGURE 11-16: Drilling the holes for the pocket

board, we can burn the main image of the moon and moon maidens. This operation took over 45 minutes to complete. The length of time varies with the associated parameters. I am something of a worrywart, so I really try to tweak parameters to get the shortest run I can be happy with. One of the techniques I used was a line density setting of 4 lines/mm to reduce the resolution. The resulting image has a small amount of *aliasing* (sometimes referred to as *stair-stepping*). Having a reasonably powerful laser (15 W) allowed me to crank the speed setting up to 5000 mm/min, which also helped.

LED diode lasers can overheat if run too long, so understanding the characteristics of your laser is critical to keep it running effectively. Heat is the primary problem and can be addressed to a degree with fans. But the best solution is to either run the laser at reduced power or limit the run time. If you have a low-powered laser and cannot run it at reduced power, break your design up into sections and burn them separately. I've even been successful at hitting pause on the CNC every 10 minutes or so and giving the laser 10–15 minutes to cool down.

With the main body correctly oriented, execute the burn pattern. (See Figure 11-17.)

The side panels work the same way. (See Figure 11-18.)

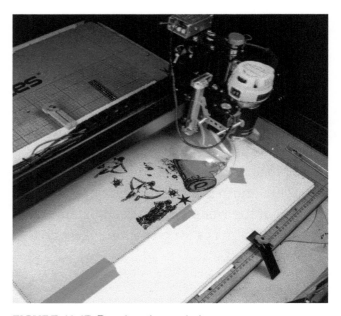

FIGURE 11-17: Burning the main image

What If I Don't Have a Laser?

While many of the operations described in this project (and the book as a whole) utilize a laser for burning operations, you can almost always achieve similar results with some other digital fabrication method. If you have a large image and no way to laser-burn it, consider printing it as a bas-relief model and pressing the leather into it. Or use your vinyl cutter and make a dye mask. Or use a CNC-controlled router to cut or engrave the image. Don't get caught up in whether you have a specific tool. Use the tools you have to create new and interesting results.

FIGURE 11-18: Burning a side panel

FIGURE 11-19: Completed outer pieces

The resulting pieces are ready for dyeing and sewing. (See Figure 11-19.)

The only other burning operation left to do is to burn the name of the movie onto the strap. The strap is 1" wide, and I'll describe its construction in the leatherworking section. The length is arbitrary and depends on the size of the wearer, but it is adjustable to a degree. There is a long section (42") and a short section (14").

We will burn the name on the long section, and this is simple to execute once the strap is oriented on the table. (See Figure 11-20.)

FIGURE 11-20: Burning the strap

LEATHERWORKING

I'm using a straightforward process to include an internal pocket in this project. I'm simply sandwiching it in between the outer pieces of leather as I stitch them. This approach is well-suited to digital leatherworking, because you can be sure that the holes will line up and be large enough. When I've done this using traditional methods, using an awl or fork to punch the holes, there is a limit on how well this works. Trying to get through too many layers of leather can be a huge pain. The CNC-drilled holes really make a difference in this kind of work.

MAKING A STRAP

Note: Cut out the strap and dye or finish it before riveting and assembly.

A bag strap differs from a belt in that it isn't a continuous loop when connected. While most people still prefer to have a buckle of some sort that allows the length of the strap to be adjusted, the two pieces that join at that buckle are individually attached to the sides of the bag. This attachment can be permanent or allow for disconnection. While I used to make permanent connections, over the years, I've come to use spring clips of one sort or another that allow me to remove or replace the strap. For this project, I used my favorite spring clip.

Attaching these clips to the ends of the strap is easy: simply slide the strap though the ring and rivet or stitch it in place. You may find it helps to cut a groove to make a sharper fold around the clip's ring. (See Figure 11-21.)

Bending the strap through the clip makes a pretty bulgy curve. To make a smoother looking spring clip–strap end, a couple of things need to happen. The first step is to cut a groove at the bend point to allow a sharper bend. (See Figure 11-22.)

The next thing that makes a big difference is to skive the folded end and cut the

FIGURE 11-21: Spring clip

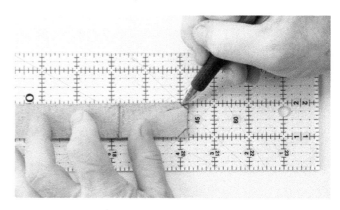

FIGURE 11-22: Strap groove and marked corners

corners. This reduces the thickness of the section and allows it to lie on the back of the strap. (See Figure 11-23.) The difference in thickness is noticeable. (See Figure 11-24.)

Two rivet holes provide a means to connect the looped strap end. Dipping the bend in water also allows the bend to be made tighter. (See Figure 11-25.)

Based on my height, I wanted a total strap length of approximately 50". As noted, I made the long end of the strap 42" and the short end 14". To create the connection for the end clips, 2" of each of these pieces were folded over. The buckle that connects the two pieces will allow some adjustment of the length.

For both aesthetics and ease of use, I trimmed the end of the short section into a gentle point so that it would slide easily through the buckle. There are English strap end punches that do this very nicely, but I just cut it with scissors and used a sanding sponge to smooth it out. (See Figure 11-26.)

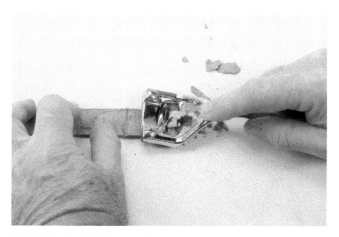

FIGURE 11-23: Skiving the strap

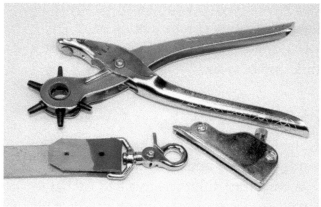

FIGURE 11-25: Strap ready for riveting

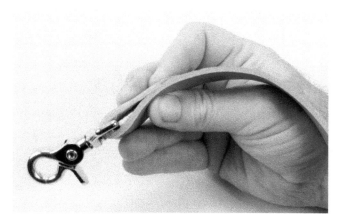

FIGURE 11-24: Skived end against the strap

FIGURE 11-26: Trimming the point on the short end of the strap

Fitting a belt strap into a buckle is easier if the strap end is slightly thinned. Once again, skiving to the rescue! In the previous skiving operation, I used a skiving tool with a razor blade. For the buckle end, I wanted to show off a more traditional skiving tool: the head knife. The scarily sharp curved blade, once you're comfortable with it, is a great tool for many operations, especially skiving. (See Figure 11-27.)

I also used an edge trimmer to take the hard edge off of the bottom of the strap. This will make the strap more comfortable to wear. (See Figure 11-28.)

While there are dozens, if not hundreds, of types of buckles to choose from, I've really come to like the Conway buckle. (See Figure 11-29.)

This buckle is a simple mechanism that doesn't have to be permanently attached to either end of the strap. I punched a hole 1" from the end of the long section of strap and threaded it through first, then I made a series of holes on the short section and threaded it through. (See Figure 11-30.)

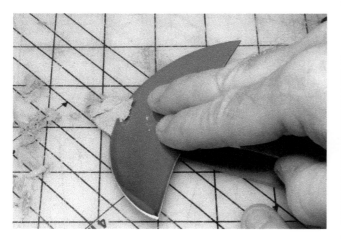

FIGURE 11-27: Skiving the belt buckle end

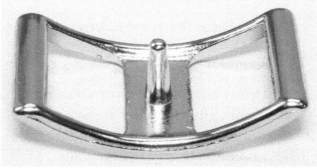

FIGURE 11-29: Conway buckle

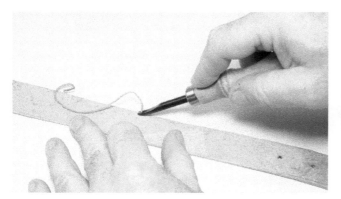

FIGURE 11-28: Beveling the strap edge

FIGURE 11-30: Using the Conway buckle

It's also easy to purchase a premade belt blank that can be used with nearly any appropriately sized buckle. (See Figure 11-31.)

COLOR AND FINISH

With the exception of the pigskin pocket, which needs no dye or finish, the steps are identical to previous projects for dyeing and finishing. Be sure to fully rinse away the carbon from the laser burned sections, then apply your favorite dye and finish combination to the main body, sides, and strap. (See Figure 11-32.)

STITCHING AND RIVETING THE BAG

Before starting to stitch the bag itself, stitch the edges of the pocket. Our holes were cut to allow these edges to fold over. Be gentle when stitching this material. If your stitches are too tight, you'll create a warp along the edge. This doesn't hurt anything, but it's less desirable than a straight edge along the front and back of the pocket. (See Figure 11-33.)

With the front and back of the pocket folded and stitched, line the holes up between the side piece, the doubled pocket, and the back of the main body. Truthfully, it's a matter of preference whether the pocket is at the front or back of the bag; just make it the same on both sides. I find that it helps to put a couple of spare needles through a couple of the holes to help keep everything aligned. (See Figure 11-34.)

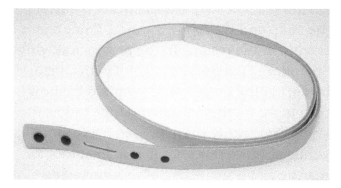

FIGURE 11-31: Premade belt blank

FIGURE 11-32: Dyeing the main body

FIGURE 11-33: Stitching the pocket edge

Stitch the four pieces together and continue around the bottom and front to fully stitch the side panel to the main body. Do this for both sides. It doesn't really matter whether you start at the front or back. I started at the front on one side and the back on another to try both. Don't forget to stitch along the front of the bag opening to reinforce the edge. (See Figure 11-35.)

The top section at the back of the bag is where we'll attach the bag to the sides with three Rapid Rivets. But don't be too hasty, because this is also where we'll attach D-rings so that we have something to connect the strap to.

First, use the holes on the main body as guides to punch holes on the adjacent section of the sides. You can go ahead and rivet the two lower holes. (See Figure 11-36.)

To create mount points for the strap clips, I cut two 2" × ½" sections of leather (don't forget to dye and finish these) and folded them each through a ½" D-ring. I held this flush against the edge of the top of the side and used the top rivet hole as a guide to punch an aligning hole. I then punched another hole lower on the D-ring strap so that it was more or less centered on the point of the side panel. Rivets went through both of these holes. The second hole allows any strain from the strap to pull on the side panel directly instead of pulling the main panel away from the side. (See Figure 11-37.)

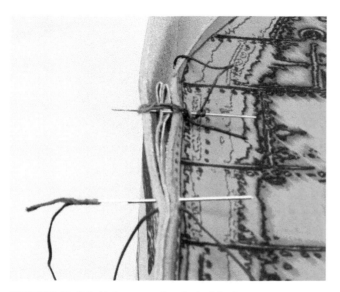

FIGURE 11-34: Using needles to hold alignment

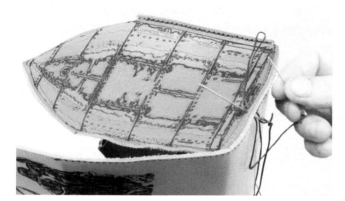

FIGURE 11-35: Stitching from the front to the back

FIGURE 11-36: Riveting the holes on the upper section

FIGURE 11-37: Punching the second hole to mount the D-rings

ATTACHING THE CLOSURE STUD

There are an unlimited number of types of closures to close the main flap of the bag, including hooks, buttons, and snaps, among others. I chose a stud. This is simple, durable, and easy to use. I first punched a button hole into the lower center of the flap. A *button hole* is a circle with a line pointing down. The hole should be the width of the thickness of the neck of the stud. The line allows the larger ball of the stud to pass through, but not easily pull out by itself. (See Figure 11-38.)

With the button hole as a guide, I used an awl to pierce the main body of the bag. The back section of the stud is pushed through this hole, and the body of the stud is threaded onto it. (See Figure 11-39.)

FIGURE 11-38: Punching the button hole

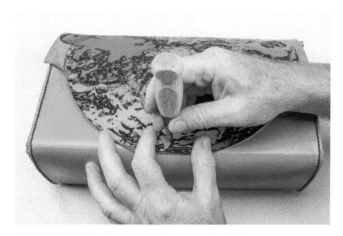

FIGURE 11-39: Making a hole to attach the stud

THE RESULTS

The resulting bag is a unique homage to a classic piece of sci-fi history! (See Figures 11-40 through 11-43.)

FIGURE 11-40: Bag front

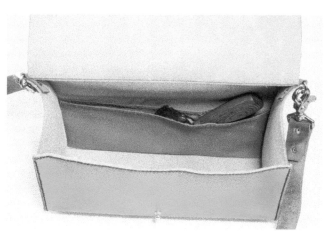

FIGURE 11-42: Bag interior

FIGURE 11-41: Bag back

FIGURE 11-43: Bag side

APPENDIX: ONLINE RESOURCES

LEATHER

Tandy Leather
tandyleatherfactory.com
Online vendor of leather and leatherworking tools

Leatherworker.net
leatherworker.net
Essential online leatherworking community and forum

Weaver Leather Supply
weaverleathersupply.com
Online vendor of leather and leatherworking tools

Zack White Leather Co.
zackwhite.com
Online vendor of leather and leatherworking tools

Fine Leatherworking
fineleatherworking.com
Online vendor of leather and leatherworking tools

The Leathercraft Guild
theleathercraftguild.com
Supporting leatherwork since 1949

HARDWARE

X-Carve CNC
inventables.com
X-Carve CNC manufacturer

Handibot
shopbottools.com
Handibot manufacturer

Prusa i3 MK2
prusa3d.com
Prusa i3 MK2 manufacturer

Printrbot
printrbot.com
Printrbot 3D printer manufacturer

2.5 W laser engraving machine
gearbest.com
Online supplier of tools and import items (2.5 W laser)

15 W TTL-enabled Focusing Laser
aliexpress.com
Online supplier of tools and import items (15 W laser)

Donek Drag Knife
donektools.com
CNC drag knife

Silhouette Stencil Cutter
silhouetteamerica.com
Vinyl and paper stencil cutters

ShopBot
shopbottools.com
CNC routers

Shapeoko
http://carbide3d.com/shapeoko/
CNC routers

Cricut
cricut.com
Vinyl and paper stencil cutters

SOFTWARE

Slic3r
slic3r.org
Slic3r slicing program

Pronterface
pronterface.com
3D printer control program

Fusion 360
http://www.autodesk.com/products/fusion-360/students-teachers-educators
3D modeling and simulation

OpenSCAD
openscad.org
3D modeling software

T2Laser
t2laser.t2graphics.com
GRBL laser engraver control software

TinkerCAD
tinkercad.com
3D modeling software

Universal Gcode Sender
https://winder.github.io/ugs_website/
GRBL G-code sending program

G-Code Q'n'dirty toolpath simulator
https://nraynaud.github.io/webgcode/
G-code simulation

Inkscape
inkscape.org
Vector drawing program

Cura
https://ultimaker.com/en/products/cura-software
3D slicing software

Repetier
repetier.com
3D slicing and control software

KISSlicer
kisslicer.com
3D slicing software

Meshmixer
meshmixer.com
Mesh creation and editing

Rhino
rhino3d.com
High-end 3D modeling using boundary definition

Adobe Creative Suite
Adobe.com
Industry-standard drawing, photography, and video tools

DIGITAL FABRICATION

Thingiverse
thingiverse.com
Vast collection of 3D models

All3DP
all3dp.com
Fantastic online 3D printing magazine

CNC Cookbook
cnccookbook.com
Fantastic CNC blog and information

Make:
makezine.com
All-around Maker resources

INDEX

Numbers

2D and 3D machining, 34–35
2-flute straight end mill, 157
2.5 W laser engraving machine, 212
3D model, 35
3D printed tools, designing, 59–60
3D printers, 32
3D printing. *See also* printing
    additive manufacturing techniques, 44–48
    arbor press, 57
    bevel cutter, 60
    blades, 59–60
    controller software, 37
    holsters, 58
    leather cases, 58
    leatherworking tools, 56–61
    molds, 58–59
    organizational tools, 60–61
    post-processing coatings, 58
    powders, 44
    "slicer" program, 36
    stamps, 56–58
    toolchain, 48–55
3DBenchy, 45
8-bit cell phone belt case
    1/18" ball end mill, 137
    3D printed frame, 132–133
    alien pattern insets, 133
    attaching spring clip, 147–149
    back flap, 134
    clean up and trim, 145–146
    clip cover, 135
    color and finish, 146–147
    design, 130–131
    digital laser fabrication, 136–145
    double-sided tape, 139
    dye beading on vinyl, 147
    finishing, 152–153
    frame model, 144–145
    front, 133–134, 152
    gap around phone, 132
    inside and back, 153
    knives, 142
    leather held for milling, 139
    leather pattern, 133–136
    leather sections, 133
    masking and spraying adhesive, 139
    midline on frame, 132
    phone dimensions, 132
    pieces laid out for cutting, 136
    prepping folds, 146
    rivets on clip cover, 148
    sewing closures, 149–150
    sewing covers, 150–151
    Silhouette model, 141
    stitching hole order, 149
    stitching through frame, 131
    taking measurements, 131–132
    VCarve simulation preview, 138
15 W TTL-enabled focusing laser, 212

A

ABS filament, 46
action cam, 112
additive manufacturing, 32, 44
Adobe Creative Suite software, 213
airbrush kit, 83–84
aliasing, 201
alien pattern
    insets, 133
    stencil, 141–144
alignment, holding, 207
All3DP digital fabrication, 214
Altoids® tin, 176
antiquing gels, 82
anvils, 105
arbor press, 57
artwork, creating, 180
ASA filament, 47
Autodesk® Fusion 360™, 35, 50, 68
Autodesk TinkerCAD, 49–50, 213
awls, 15–16

B

back-stitching, 28
belly of hide, 4
belt pouch
    clean up and trim, 187–188
    color and finish, 188–189
    creating Cthulhu, 180–181
    cutting away fibers, 188
    digital laser fabrication, 182–187
    finishing, 190–191
    grooves cut, 186
    holes and outline, 185
    image toolpath, 184
    leather workpiece, 184
    pattern, 176–180
    preparing files, 182–187
    setting snaps, 189
    stitching, 189
    taking measurements, 176
    X0Y0, 187
bend of hide, 4
beveling
    and edging, 14
    tools, 59, 60–61
blades, 59–60
boosh controller PCB, 157–158
boundary definition, 52–53
box cutters, 10
brain tanning, 2
break-in, accelerating, 85–86
BS (acrylonitrile butadiene styrene), 46
buckles and rings, 19–20
burn pattern, multi-tool holder, 96
burning
    and engraving, 169–170
    operations, 200–202
button hole, punching, 208
buttons and studs, 20–21

C

CAD (computer assisted design), 34–35. *See also* sheet metal CAD
Cahier, 63–64
calipers, 112
camera model, 112, 114, 119
carbon fiber filament, 46
carbon stains, avoiding, 80
cardboard check, running, 78
Cartesian machine, 33
carving and tooling, 23
cell phone belt case
    1/18" ball end mill, 137
    3D printed frame, 132–133
    alien pattern insets, 133
    attaching spring clip, 147–149
    back flap, 134
    clean up and trim, 145–146
    clip cover, 135
    color and finish, 146–147
    design, 130–131
    digital laser fabrication, 136–145
    double-sided tape, 139

dye beading on vinyl, 147
finishing, 152–153
frame model, 144–145
front, 133–134, 152
gap around phone, 132
inside and back, 153
knives, 142
leather held for milling, 139
leather pattern, 133–136
leather sections, 133
masking and spraying adhesive, 139
midline on frame, 132
phone dimensions, 132
pieces laid out for cutting, 136
prepping folds, 146
rivets on clip cover, 148
sewing closures, 149–150
sewing covers, 150–151
Silhouette model, 141
stitching hole order, 149
stitching through frame, 131
taking measurements, 131–132
VCarve simulation preview, 138
channels, cutting, 81–82
Chartres Cathedral labyrinth, 65
chisels, 16
chrome tanning, 2, 5
clasps and hooks, 20
cleaning filament, 47
clips, using to mold leather, 23
closure stud, attaching for shoulder bag, 208
closures
    fractal journal cover, 87
    types, 19–21
CNC (computer numeric control). *See also* G-code
    Cartesian machine, 33
    router, 33
    toolpath, 36
CNC Cookbook, 214
CNC drag knife, 142
CNC operations
    materials, 39
    tablet sleeve, 166–170
CNC swivel knife, 12
collagen fibrils and fibers, 1–2
color and finish
    cell phone belt case, 146–147
    fractal journal cover, 82–84
    multi-tool holder, 102–104
color changing filament, 47
conchos and spots, 22
conditioners, 86
conductive filament, 47
constructing images and models, 34–35
controllers, 36–37, 55
copper layers, 158–159
copper rivets and washers, 18
corium, 2
corners, cutting, 86–87
Cricut hardware, 212
CSG vs. mesh model, 58–59
Cthulhu, 175, 180–181, 183–184, 187
Cura software, 213
curing resin, 45
curved lines, measuring, 197
cutting
    corners, 86–87
    fold groove, 81
    leather, 9–14
    multi-tool holder, 101
    stitching groove, 82
    threads, 29

D

decorating leather, 22–29
digital calipers, 112
digital imagery, multi-tool holder, 91–92
digital laser fabrication
    3D printed parts, 163–164
    attaching spring clip, 147–148
    cardboard check, 78
    clean up and trim, 80, 101–102, 119–121, 145–146
    CNC operations, 166–170

CNC work, 136–137, 184–186
color and finish, 121–124, 146–147
contact-paper layer, 167
cover burn plan, 70–71
cover cut plan, 69
cutting, 76–81, 101
cutting consistency, 76
cutting out pieces, 139
cutting pattern stencil, 141–144
dimensions, 70
DXF (Drawing Exchange Format), 76, 94
DXF editor, 84
engraving, 79–81, 101
engraving and burning, 169–170
explained, 32–33
extruding, 94–95
focusing laser, 76
G-code, 72
homing machines, 71
images, 163
laser cutting and burning, 117–118
laser work, 187
layout guide, 74
leather shrinkage, 77
machine setup, 71, 73–74
making gears, 118–119
making rough cuts, 75
manual position setting, 73
manual stitching-holds, 68
milling and drilling, 168–169
molding leather, 164–165
mounting gears, 126
optimized toolpath, 69
orienting workpiece, 167
pocket cut plan, 70
pocket piece, 75
preparing files, 67–70, 93–101, 116, 182–184
prepping folds, 146
prepping workpiece, 74–76, 100–101, 116–117, 184
prepping workpieces, 137–139
printing camera model, 119
printing frame, 144–145
    pulsing and jogging laser, 74
    resources, 214
    riveting gears, 125–126
    setting eyelets, 125–126
    setting machine origin, 73
    sewing closures, 149–150
    sewing corners, 124–125
    sewing covers, 150–151
    sleeve parts, 161–163
    steampunk action-cam top hat, 116–119
    tablet sleeve, 161–170
    toolpath spaghetti, 68
    unfolding, 179
    workholding sandwich, 167
    x-carriage, 73
    x-gantry, 73
digital leatherworking, 39–41
dimensions, handling, 70, 99
Donek drag knife, 212
double bend of hide, 4
double shoulder of hide, 4
drive punch, 13
DXF (Drawing Exchange Format), 35, 68, 76
DXF editor, using, 84
DXF files, steampunk action-cam top hat, 116
dyeing leather, 82–83, 103, 122, 147, 168, 206

E

Easel CAD/CAM tool, 136
edge roller, 172
edging and beveling, 14
elder gods belt pouch. *See* pouch
ELE Explorer action cam, 112
embroidery scissors, 9
engraving
    and burning, 169–170
    fractal journal cover, 79–80
    multi-tool holder, 101
    with shades of gray, 194
extruding, 94–95
eyelets, 17–18
    setting, 125–126

F

fibers
    cutting away, 188
    filing, 145
filament characteristics, 46–48
fillet, measuring, 160
Fine Leatherworking, 211
finish and color, 82–84, 102–104
flesh, 2
flexible TPE/TPU filament, 46
fold groove, cutting, 81
folds, prepping, 146
forked chisels, 16
FPE filament, 48
fractal journal cover
    accelerating break-in, 85–86
    Chartres Cathedral labyrinth, 65
    cleanup and trim, 80
    closure, 87
    closures, 87
    color and finish, 82–84
    cover pattern, 66
    cutting, 76–81
    cutting channels, 81–82
    cutting corners, 86–87
    digital imagery, 65–66
    digital laser fabrication, 67
    engraving, 76–81
    extending, 87
    finishing, 85
    fold allowance, 64
    G-code, 72
    Julia set, 66
    leather pieces, 65
    machine setup, 71, 73–74
    Mandeltree, 65
    pattern, 64–65
    preparing files, 67–70
    prepping workpiece, 74–76
    raster images, 67
    results, 85
    snaps and studs, 87
    stitching, 84–85
frame model, 144–145
fritzing layer options, 158
full grain leather, 3
Fusion 360 software, 35, 50, 68, 213

G

G-code. *See also* CNC (computer numeric control)
    digital laser fabrication, 72
    features, 36–37
    simulator, 68
G-Code Q'n'dirty toolpath simulator, 213
gear model, 114
gears
    making for steampunk top hat, 118–119
    mounting, 126
    painting, 124
    riveting, 125–126
genuine leather, 3
Gerber files, 158
GIMP (GNU Image Manipulation Program), 34, 194–195
glow-in-the-dark filament, 47
glue, 21
gouging and grooving, 13
grains of leather, 2–3
gray levels, 194–195
grommets, 17–19
grooving and gouging, 13

H

hammering, 105
hammering metal, 13
hand punch, 16. *See also* punching
Handibot hardware, 212
hand-stitching, 14–15, 24–29
hatband model, 113. *See also* steampunk action-cam top hat
head knife, 10
hides, 3–5
HIPS filament, 46
hole order, stitching, 149

holes
- even patterns, 15–16
- getting needles through, 28
- making, 12
- punching, 26
- shrinkage, 80

holsters, 58
homing machines, 71
hooks and clasps, 20
hydrophilicity, 40

I

image, engraving with shades of gray, 194
image, constructing, 34–35
Inkscape program, 34, 180, 213

J

joining leather, 14–21
journal cover
- accelerating break-in, 85–86
- Chartres Cathedral labyrinth, 65
- cleanup and trim, 80
- closure, 87
- closures, 87
- color and finish, 82–84
- cover pattern, 66
- cutting, 76–81
- cutting channels, 81–82
- cutting corners, 86–87
- digital imagery, 65–66
- digital laser fabrication, 67
- engraving, 76–81
- extending, 87
- finishing, 85
- fold allowance, 64
- G-code, 72
- Julia set, 66
- leather pieces, 65
- machine setup, 71, 73–74
- Mandeltree, 65
- pattern, 64–65
- preparing files, 67–70
- prepping workpiece, 74–76
- raster images, 67
- results, 85
- snaps and studs, 87
- stitching, 84–85

Julia set, fractal journal cover, 66

K

KISSlicer software, 213
knives
- CNC drag vs. stencil cutter, 142
- and scissors, 9–10

L

laser alternatives, 182, 201
laser fabrication. *See* digital laser fabrication
laser pattern, aligning, 200
laser sintering, 44
laser tool, using, 71, 74–75
leather
- conditioners, 86
- cutting, 9–14
- dyeing, 82–83
- grades, 2
- holding while sewing, 27
- hydrophilicity, 40–41
- inconsistencies, 41
- irregularities, 92
- joining, 14–21
- molding, 40
- properties, 39–40
- quality, 4–5
- versus rawhide, 3
- recycling, 5
- removing, 9–14
- salvaging, 5
- shaping and decorating, 22–29
- shrinkage, 77
- softening, 85–86
- staining, 82
- suntanning, 41
- surprises, 41

thickness, 40
thicknesses, 3–4
weight, 4
leather cases, 58
leather pattern, multi-tool holder, 90–91. *See also* patterns
Leatherman Wave®, 90
leather-stitching needle, 17
Leatherworker.net, 211
leatherworking
    attaching closure stud, 208
    attaching spring clip, 147–148
    clean up and trim, 119–121, 145–146, 171, 187–188
    color and finish, 82–84, 102–104, 121–124, 146–147, 188, 206
    cutting channels, 81–82
    cutting corners, 86
    finish coat, 171
    making straps, 203–206
    prepping folds, 146
    riveting, 106–108, 125–126, 206–207
    setting eyelets, 125–126
    setting snaps, 104–106, 189
    sewing closures, 149–150
    sewing corners, 124–125
    sewing covers, 150–151
    stitching, 84–85, 189, 206–207
    tablet sleeve, 171–172
leatherworking tools. *See also* multi-tool holder
    beveling and edging, 14
    caution, 9
    grooving and gouging, 13
    knives and scissors, 9–10
    punches, 12–13
    resource, 8
    skiving, 11
    storing, 60
    swivel knives, 11–12
LED diode lasers, overheating, 201
lignin (bioFila) filament, 46
Line 20 snap, 19
line segments, joining, 97
liquids, printing with, 45

M

machines, homing, 71
machining modes, 34–35
magnetic filament, 47
magnetic snaps, 149. *See also* snaps and studs
magnets, 21
Maker resources, 214
Mandelbrot, Benoit, 65
Mandeltree, 65
manufacturing tools, 32
measurements
    cell phone belt case, 131–132
    steampunk action-cam top hat, 112–114
    tablet sleeve, 160
mechanical iris, 127
Méliès, Georges, 193
mesh modeling, 52–53
mesh vs. CSG model, 58–59
Meshmixer software, 213
metal filament, 46
metal on mental, caution, 13
metal stamps, 57–58
metallic powders, 44
milled leather, 139, 159
milling and drilling, 168–169
miter, splitting model with, 178–179
model, constructing, 34–35
modeling objects
    Autodesk Fusion 360™, 35, 50, 68
    Autodesk TinkercCAD, 49–50
    boundary definition, 52–53
    mesh modeling, 52–53
    overview, 58–59
    OpenSCAD, 51–52, 113, 213
molding leather, 22–23, 40
molds, 3D printing, 58–59
Moleskine Cahier, 63–64
motor, moving by steps, 38

multi-tool holder. *See also* leatherworking
tools
    applying resist, 103–104
    burn pattern, 96, 98
    burning gears, 102
    clean up and trim, 101–102
    clipped gears, 98
    cut pattern, 91
    cutting, 101
    digital imagery, 91–92
    digital laser fabrication, 93–102
    engraving, 101
    filled pattern outline, 97
    finishing, 109
    gears and skull, 99
    gears positioned, 97
    joining line segments, 97
    leather pattern, 90–91
    leatherworking, 102–104
    nested patterns, 100
    overlapping sections, 100
    preparing files, 93–100
    prepping workpiece, 100–101
    results, 109
    riveting, 106–108
    setting snap, 104–106
    skull image, 92, 94
    unjoined line segments, 96

N

NC (numeric control), 31, 72
needles
    getting through holes, 28
    for hand-stitching, 25, 27–28
    for holding alignment, 207
    using, 16–17
nGen filament, 47
notebook. *See* fractal journal cover
nylon filament, 46

O

OpenSCAD software, 51–52, 113, 213

organizational tools, 60–61
overlapping sections, 100

P

Parsons, John, 31
patterns, nesting, 100. *See also* leather pattern
PC polycarbonate, 46
PC/ABS filament, 47
PCB layers, 158
PCB tablet sleeve. *See* boosh controller PCB;
    tablet sleeve
PCB traces, milling, 170
Pen tool, 180
PET (CEP) filament, 46
PETG (XT, n-vent) filament, 46
PETT (tglase) filament, 47
pill bottle storage system, 60
Piñatex™, 3
PLA (polylactic acid), 46
plastic filament, printing with, 45–48
pleather, 3
PMMA, acrylic filament, 47
PNG format, 35
pocket edge, stitching, 206
polymerization, 45
POM, acetal filament, 47
PORO-LAY filament, 48
posterized image, 195
pouch
    clean up and trim, 187–188
    color and finish, 188–189
    creating Cthulhu, 180–181
    cutting away fibers, 188
    digital laser fabrication, 182–187
    finishing, 190–191
    grooves cut, 186
    holes and outline, 185
    image toolpath, 184
    leather workpiece, 184
    pattern, 176–180
    preparing files, 182–187
    setting snaps, 189
    stitching, 189

taking measurements, 176
X0Y0, 187
powders, printing with, 44
PP filament, 47
printing. See also 3D printing
    frame, 144–145
    with liquids, 45
    with plastic filament, 45–48
    with powders, 44
Printrbot hardware, 212
Pronterface software, 213
Prusa i3 MK2 hardware, 212
pulse train, 38–39
punching. See also hand punch
    button hole, 208
    explained, 12–13
    holes, 26
PVA filament, 46
PWM (pulse width modulation), 194–195

Q

quality of leather, 4–5

R

Rapid Rivet, 18–19
raster images, 34–35, 67
rawhide, 3
Raynaud, Nicolas, 68
recycling leather, 5
removing leather, 9–14
Repetier software, 213
RepRap G-code program, 72
resin printing, 45
resist
    applying, 103–104
    leaking, 143
resources. See websites
Rhino software, 213
rings and buckles, 19–20
riveting
    explained, 106–108
    gears, 125–126
    shoulder bag, 206–208
rivets, 17–19, 148
rotary knives, 10
rotary punch, 12
round knife, 10
router, 33
rubber stamps, 57–58

S

saddle stitch, 15, 24–29, 150. See also stitching
salvaging leather, 5
sandstone (LAYBRICK) filament, 47
scissors and knives, 9–10
screw post, 17
screws, 17–19
sections, overlapping, 100
servos, 37
setting snap, 105–106
sewing awl, 16
sewing leather, 14–17
Shapeoko hardware, 212
shaping leather, 22–29
shears, 9
sheet metal CAD, 177. See also CAD (computer assisted design)
ShopBot hardware, 212
shoulder bag
    aligning laser pattern, 200
    attaching closure stud, 208
    burning operations, 200–202
    burning strap, 202
    color and finish, 206
    cutting and drilling, 199–200
    design, 194–199
    digital laser fabrication, 199–202
    drilling holes for pocket, 200
    dyeing, 206
    finishing, 209
    Le Voyage dans la Lune, 193
    making strap, 203–206

posterized image, 195
punching button hole, 208
PWM (pulse width modulation), 194–195
stitching and riveting, 206–208
stitching pocket edge, 206
shoulders of hide, 4
side of hide, 4–5
Silhouette model, 141
Silhouette stencil cutter, 212
silkscreen layer, 159, 170
single shoulder of hide, 4
sintering, 44
skiving. *See also* thickness of hide
    explained, 11
    strap for shoulder bag, 204
skull image, 92, 94–95
Slic3r software, 213
"slicer" program, 36
slicing, 53–55
snap-blade knives, 10
snaps and studs. *See also* magnetic snaps
    setting, 189
    stitching, 150
    using, 19, 87, 104–106
sockets, setting, 106
softening leather, 85–86
split of hide, 4
spoil board, surfacing, 140–141
spots and conchos, 22
spring clip, attaching, 147–148
spring punch, 12
staining leather, 82
stair-stepping, 201
stamp punch, 13
stamping, 24, 105
stamps, 3D printing, 56–58
steampunk action-cam top hat
    #2 and #3/0 brushes, 123
    3D printed pieces, 123–124
    applying dye, 122
    clean up and trim, 119–121
    color and finish, 121–124
    cutting grooves, 120

    digital laser fabrication, 116–119
    edge trimming, 120
    enhancements, 127
    finishing, 126–127
    gears, 118–119
    groove locations for folding, 120
    leather pattern, 114–115
    leather placed for cutting, 117
    mounting gears, 125
    painting gears, 124
    painting pattern, 122
    pattern check, 117
    pattern test, 114
    resisting pattern parts, 122
    riveting gears, 125–126
    scanned border pattern, 115
    setting eyelets, 125–126
    sewing corners, 124–125
    stitching holes, 125
    stitching pattern, 125
    taking measurements, 112–113
    work blank, 117
    wrapping edge, 121
stencil cutter knife, 142
stencil pattern, cutting, 141–144
step and direction commands, 37–39
stepper motors, 37–39
stitches, pulling tight, 28
stitching. *See also* saddle stitch
    explained, 84–85
    from front to back, 207
    hole order, 149
    pattern, 125
    pocket edge, 206
    pouch, 189
    shoulder bag, 206–208
    snaps, 150
    tablet sleeve, 171
stitching groove, 13, 82
stitching holes
    clearing, 80
    marking, 125
stitching pony, 27

STL files, steampunk top hat, 116
Stohlman, Al, 8, 15, 29
strap for shoulder bag
- beveling edge, 205
- burning, 202
- Conway buckle, 205
- making, 203–206
- skiving, 204

studs
- and buttons, 20–21
- and snaps, 87, 105

subtractive manufacturing tools, 32
suede, 2
suntanning leather, 41
surfacing tables and boards, 140–141
SVG format, 35
swivel knives, 11–12, 61

T

T2Laser software, 213
tablet sleeve
- 2-flute straight end mill, 157
- art, 157–160
- boosh controller PCB, 157
- clean up and trim, 171
- contact-paper layer, 167
- copper layers, 158–159
- design, 156
- digital laser fabrication, 161–170
- edge coating, 172
- finish coat, 171
- finishing, 173
- fritzing layer options, 158
- Gerber files, 158
- milled leather, 159
- sewing and finishing, 171–173
- silkscreen layer, 159
- stitching, 171
- structure, 156–157
- taking measurements, 160
- traces, 158

Tandy Leather, 8, 211
tanning leather, 2–3
thickness of hide, 4, 40. *See also* skiving
Thingiverse digital fabrication, 214
threads
- cutting, 29
- for hand-stitching, 25–26

TinkerCAD software, 49–50, 213
toggle button, 20
toolchains, 34–39, 48–55
tooling and carving, 23
toolpaths, generating, 35–37, 136
tools. *See* leatherworking tools
top grain, 3
top hat. *See* steampunk action-cam top hat
TPC filament, 47
TPE/TPU filament, 46
traced vector files, cleaning up, 95. *See also* vector images
traces, 158, 169
trimming corners, 86–87
tubular rivets, 18, 106–107, 148

U

Universal Gcode Sender software, 213
utility knives, 10

V

VCarve simulation preview, 138
vector images, 34. *See also* traced vector files
vegetable tanning, 2–3, 40
Venturi effect, 83
Vetric VCarve Pro, 136
V-gouges, 13, 81–82
vinyl cutter swivel knife, 12

W

water, 40
wax (MOLDLAY) filament, 47
Weaver Leather Supply, 211
websites
- 2.5 W laser engraving machine, 212

15 W TTL-enabled focusing laser, 212
Adobe Creative Suite software, 213
All3DP digital fabrication, 214
CNC Cookbook, 214
Cricut hardware, 212
Cura software, 213
digital laser fabrication, 214
Donek drag knife, 212
Fine Leatherworking, 211
Fusion 360 software, 213
G-Code Q'n'dirty toolpath simulator, 213
Handibot hardware, 212
Inkscape program, 213
KISSlicer software, 213
Leatherworker.net, 211
Maker resources, 214
Meshmixer software, 213
OpenSCAD software, 213
Printrbot hardware, 212
Pronterface software, 213
Prusa i3 MK2 hardware, 212
Repetier software, 213
Rhino software, 213
Shapeoko hardware, 212
ShopBot hardware, 212
Silhouette stencil cutter, 212
Slic3r software, 213
T2Laser software, 213
Tandy Leather, 211
Thingiverse digital fabrication, 214
TinkerCAD software, 213
Universal Gcode Sender software, 213
Weaver Leather Supply, 211
X-Carve CNC hardware, 212
XTC-3D coating, 58
Zack White Leather Co., 211
weight of hide, 4
Wikimedia.org, 141
wood filament, 46
work areas, handling, 99
work table, surfacing, 140–141

X

X0Y0, determining for belt pouch, 187
x-carriage, digital laser fabrication, 73
X-Carve CNC hardware, 212
x-gantry, digital laser fabrication, 73
XTC-3D coating, 58

Z

Zack White Leather Co., 211
Z0 position, finding, 185–186
Z-probe, 185